AF735034

Bacterial Blight of Rice

Bacterial Blight of Rice offers a comprehensive and authoritative account of one of the most destructive diseases threatening global rice production. Rice (*Oryza sativa* L.), the staple food for more than half of the world's population, suffers severe yield losses—often 30–60%—from bacterial blight caused by *Xanthomonas oryzae* pv. *oryzae* (*Xoo*). Although nearly 50 resistance (*Xa*) genes have been deployed in breeding programs, the rapid evolution of *Xoo* races continues to undermine durable resistance, demanding a deeper understanding of pathogen diversity and host–pathogen interactions.

This book brings together foundational knowledge and the latest scientific advances to deliver a comprehensive account of bacterial blight. It covers disease history, symptoms, diagnostics, population biology, and pathotyping, with strong emphasis on *Xoo* genomics, *hrp* gene clusters, and effector-driven pathogenesis mediated by the type III secretion system. Cutting-edge resistance and management strategies—including gene pyramiding, genome-enabled breeding, genome editing, and eco-friendly, sustainable disease control—are presented in a clear, practical framework.

Distinctive for its systematic organization and research-backed content, the book also provides hands-on guidance for key techniques such as pathogen inoculation and virulence screening. Designed as both a textbook and a reference, it will be invaluable to students, researchers, breeders, extension professionals, policymakers, and progressive farmers.

Kalyan K. Mondal has over two decades of experience in bacterial blight research and has led multiple funded projects on *Xoo* type III secretion system effectors. His translational research enabled the development of seven bacterial blight–resistant rice varieties, significantly strengthening rice productivity. He also serves as an expert reviewer for the EuroXanth DokuWiki under the European COST program.

Bacterial Blight of Rice

Kalyan K. Mondal

CRC Press
Taylor & Francis Group
Boca Raton London New York

CRC Press is an imprint of the
Taylor & Francis Group, an **informa** business

First edition published 2027
by CRC Press
2385 NW Executive Center Drive, Suite 320, Boca Raton FL 33431

and by CRC Press
4 Park Square, Milton Park, Abingdon, Oxon, OX14 4RN

CRC Press is an imprint of Taylor & Francis Group, LLC

© 2027 Kalyan K. Mondal

Reasonable efforts have been made to publish reliable data and information, but the author and publisher cannot assume responsibility for the validity of all materials or the consequences of their use. The authors and publishers have attempted to trace the copyright holders of all material reproduced in this publication and apologize to copyright holders if permission to publish in this form has not been obtained. If any copyright material has not been acknowledged please write and let us know so we may rectify in any future reprint.

Except as permitted under U.S. Copyright Law, no part of this book may be reprinted, reproduced, transmitted, or utilized in any form by any electronic, mechanical, or other means, now known or hereafter invented, including photocopying, microfilming, and recording, or in any information storage or retrieval system, without written permission from the publishers.

For permission to photocopy or use material electronically from this work, access www.copyright.com or contact the Copyright Clearance Center, Inc. (CCC), 222 Rosewood Drive, Danvers, MA 01923, 978-750-8400. For works that are not available on CCC please contact mpkbookspermissions@tandf.co.uk

For Product Safety Concerns and Information please contact our EU representative GPSR@taylorandfrancis.com. Taylor & Francis Verlag GmbH, Kaufingerstraße 24, 80331 München, Germany.

Trademark notice: Product or corporate names may be trademarks or registered trademarks and are used only for identification and explanation without intent to infringe.

ISBN: 978-1-041-20008-6 (hbk)
ISBN: 978-1-041-20009-3 (pbk)
ISBN: 978-1-003-71459-0 (ebk)

DOI: 10.1201/9781003714590

Typeset in Times
by Apex CoVantage LLC

Contents

Foreword

Trust for Advancement of Agricultural Sciences
Avenue II, Indian Agricultural Research Institute, New Delhi - 110 012
Phone : 011-25843243; +91-8130111237
E-mail : taasiari@gmail.com Website : www.taas.in
taas
Progress Through Science

Chairman
Dr RS Paroda
Vice-Chairman
Dr T Mohapatra
Secretary
Dr Bhag Mal
Treasurer
Dr JL Karihaloo
Trustees
Dr ML Jat
Dr Renu Swarup
Dr BS Dhillon
Dr Ch Srinivas Rao
Dr Raju Barwale

Foreword

Rice (*Oryza sativa* L.) is the foundation of life and livelihood for billions across the globe. With Asia alone contributing more than 90% of the world's rice, its sustained prosperity is vital for stability and meeting nutritional needs. For millions of farmers, particularly in countries like India, rice is the backbone of their income, heritage, and cultural identity. Despite the global importance, rice production faces relentless challenges. Among these, bacterial blight (BB), caused by *Xanthomonas oryzae* pv. *oryzae* (Xoo), remains one of the most destructive diseases. Severe outbreaks can decimate entire harvests and devastate the livelihoods of smallholder farmers and poses a socio-economic challenge. For decades, the central management strategy has been developing resistant rice through the pyramiding of resistance genes (Xa genes). While this provided significant relief, the adaptability and genetic diversity of Xoo often overcome resistance. This ongoing "arms race" underscores the urgent need for sustained innovative research and modern molecular tools.

Encouragingly, recent advances in plant genomics, molecular biology, and biotechnology have opened promising new avenues. A deeper understanding of pathogen effectors, such as transcription activator-like effectors (TALEs), and their host targets, like the *OsSWEET* genes, has revolutionized our ability to design precise resistance strategies. Natural allelic variants such as *xa13*, *xa25*, and *xa41*, coupled with cutting-edge tools like CRISPR–Cas genome editing, demonstrate how pathogen strategies can be effectively blocked. These scientific breakthroughs hold transformative potential for creating broad-spectrum, durable resistance that can reach to farmers' fields.

The book *Bacterial Blight of Rice*, published by CRC Press, is a significant contribution because it bridges fundamental science with practical application. It consolidates decades of research on host–pathogen interactions, molecular mechanisms, and resistance breeding, while keeping the farmer's perspective at its core. It highlights how knowledge and innovation transform into effective solutions as the future of the food security of billions depends on how effectively we translate scientific knowledge into durable, farmer-friendly solutions.

I congratulate Dr. Kalyan K. Mondal, for his over two decades of rigorous research conducted under the mentorship of the renowned bacteriologist, the late Prof. Jeevan Prakash Verma. This book will undoubtedly serve as an invaluable resource for students, researchers, breeders, and academicians alike—both as a repository of scientific knowledge and as a source of inspiration.

Dr. R.S. Paroda
President, IARI Alumni Association (IAA)
Former Secretary, DARE & DG, ICAR, New Delhi

Preface

Rice (*Oryza sativa* L.) is one of the world's most important staple crops, forming the basis of numerous cuisines worldwide. To sustain global food security amid the rapidly growing population—particularly in Asia—it is essential to enhance rice productivity despite challenges posed by both biotic and abiotic stresses. Among the biotic threats, bacterial blight (BB), caused by *Xanthomonas oryzae* pv. *oryzae* (*Xoo*), is one of the most destructive diseases, significantly reducing rice yields across continents. In India, as in other rice-growing Asian countries, rice-based farming systems dominate, and the livelihoods of millions of farmers revolve around rice cultivation. However, BB continues to appear with varying severity, leading to yield losses estimated between 30% and 60%.

The disease manifests in distinct forms depending on the growth stage of the crop. At the seedling stage, or shortly after transplanting, BB often causes the "kresek" phase, characterized by sudden yellowing and wilting of seedlings within a week of transplanting. The affected seedlings eventually dry out completely, resulting in total crop failure in the field. At later stages, particularly during active tillering, the more common "leaf blight" phase occurs. This begins as water-soaked lesions on leaf tips or margins, progressing downward or inward in a wavy pattern, eventually leading to complete blighting of the leaf blade. Severely affected fields display a scorched, blighted appearance when viewed from a distance.

The most sustainable approach to BB management is the deployment of resistant cultivars through pyramiding of resistance (*R*) genes into high-yielding rice varieties. To date, nearly 50 *R* genes (*Xa* genes) against BB have been identified and utilized in breeding programs. However, resistance in such varieties is often short-lived due to the rapid evolution of *Xoo* races that can overcome or bypass *R* gene-mediated defense. This underscores the need for continuous monitoring of *Xoo* virulence diversity and the incorporation of characterized pathogen populations in breeding strategies. In India, *Xoo* populations have been studied extensively with respect to their virulence profiles, genome structures, transcriptomes, and type III secretion system (T3SS) effector repertoires.

Xoo delivers two major classes of effector proteins into rice cells through its T3SS: transcription activator-like effectors (TALEs) and *Xanthomonas* outer proteins (Xops). These effectors suppress host immune responses, including both PAMP-triggered immunity (PTI) and effector-triggered immunity (ETI). Genome analyses of *Xoo* race 4, a predominant race in India, identified 18 TALEs and 21 Xop effectors. TALEs function by directly binding to effector-binding elements (EBEs) in the promoters of host susceptibility (*S*) genes, such as *OsSWEET* genes, thereby activating their expression and promoting pathogen nutrition. Of the five *OsSWEET* genes, three (*OsSWEET11*, *OsSWEET13*, and *OsSWEET14*) are natural TALE targets. Specific TALEs such as PthXo1, PthXo2, PthXo3, AvrXa7, TalC, and TalF target different EBEs in these genes. Notably, *OsSWEET12* and *OsSWEET15* can be artificially induced but have not yet been shown to be targeted by natural *Xoo* strains.

DNA polymorphisms in EBEs can disrupt TALE binding and block pathogen-induced susceptibility. For example, the naturally occurring promoter mutation in *OsSWEET11* (known as *xa13*) confers resistance to *Xoo* without yield penalties and has been widely used in breeding programs. Similarly, recessive resistance alleles *xa25* (*OsSWEET13*) and *xa41* (*OsSWEET14*) provide durable resistance. Building on these insights, gene-editing approaches targeting single or multiple EBEs in *SWEET* promoters have been developed, showing strong promise for generating broad and durable resistance against major *Xoo* races.

Non-TALE effectors (Xops) also play crucial roles in BB pathogenesis. Unlike TALEs, Xop proteins are localized in the plant cytoplasm, where they interact with host proteins to suppress immunity and promote bacterial survival. Many Xop effectors have been shown to interfere with PTI by modulating defense responses such as callose deposition, reactive oxygen species (ROS) accumulation, and the expression of defense-related genes. The functional diversity of over 20 identified Xop effectors highlights their importance in disease development.

This book aims to provide an updated and comprehensive understanding of BB, focusing on both the pathogen and its interaction with rice. It covers the distribution, epidemiology, symptomatology (with field images), molecular basis of pathogenesis, effector repertoires, and current management strategies. Emphasis is placed on recent advances in *Xoo* population biology, genomics, and effector research, along with host resistance mechanisms and novel disease management options, including genome editing. The content is designed to serve researchers, students, extension specialists, policymakers, and rice farmers alike. Altogether, *Bacterial Blight of Rice* is intended as both a reference and a textbook, equipping stakeholders with the knowledge base required to advance future research and practical solutions for managing this devastating disease.

I deeply acknowledge the unwavering support and encouragement extended by my mother, my late father, and all my family members throughout the process of writing this book. Their love, sacrifices, and constant motivation have been a source of strength for me at every stage. I also express my heartfelt gratitude to all my teachers, whose wisdom, guidance, and inspiration have continually shaped my thoughts and motivated me during the compilation of this work.

About the Author

Dr. Kalyan K. Mondal, an eminent plant pathologist, has made pioneering contributions to the understanding and management of bacterial diseases of crops. He earned his B.Sc. (Ag.) Hons. and M.Sc. (Ag.) in Plant Pathology from Bidhan Chandra Krishi Viswa Vidyalaya, West Bengal, in 1991 and 1993, respectively, followed by a Ph.D. in Plant Pathology from ICAR-IARI, New Delhi, in 1999. After beginning his scientific career as Scientist at CSKHPKV, Palampur (2001–2006), he joined ICAR-IARI, New Delhi, where he served as Senior Scientist (2006–2012) and Principal Scientist (2012–2023). Since August 2023, he has been serving as Joint Director (Research) at ICAR-National Institute of Biotic Stress Management, Raipur, Chhattisgarh.

As a Visiting Scholar at Stanford University, California, USA, Dr. Mondal received advanced training on the bacterial type-III secretion system (T3SS) and its role in subverting plant defense. He is credited with the first comprehensive elucidation of the effector repertoire of *Xanthomonas oryzae* pv. *oryzae* (*Xoo*) in rice and *X. axonopodis* pv. *punicae* (Xap) in pomegranate, and with functionally establishing the critical role of these effectors in disease development. He also generated the first complete genome sequence of *Xoo* race 4, a virulent Indian strain of global importance.

Dr. Mondal's translational research has had far-reaching impact on Indian agriculture. He was actively involved in the development and commercialization of seven bacterial blight–resistant rice varieties—Pusa 1592, Pusa 1718, Pusa 1728, Pusa 1847, Pusa 1885, Pusa 1886, and Samba 1853—harboring the resistance genes *xa13* and *Xa21*, which have significantly enhanced productivity and resilience in rice cultivation.

For his outstanding contributions, Dr. Mondal has been elected Fellow of NAAS (2023), NABS (2020), ARRW (2021), IPS (2013), and SARP (2020), and is also a Member of NASI (2013). His research excellence has been recognized with numerous national awards, including the Prof. S. Kannaiyan Memorial NABS-Lifetime Achievement Award (2025), Nanaji Deshmukh ICAR Award (2021), Dr. B.P. Pal Memorial NABS Best Scientist Award (2022), Prof. J.P. Verma Memorial Award (2016), Prof. M.K. Patel Memorial Young Scientist Award (2008), and the Prof. M.J. Narasimhan Academic Merit Award (2001).

Beyond research, Dr. Mondal has made impactful contributions to scientific leadership and policy through his roles in national task forces, project review panels, and editorial boards. He has served as Task Force Member (DBT, Agricultural Biotechnology), Project Reviewer (DST-SERB), Member of Institute Management Committee (ICAR-National Bureau of Agriculturally Important Microorganisms, 2016–2019), Chief Editor (2026–2028), Editor (2017–2019) of *Indian Phytopathology*, Joint Secretary of the Indian Phytopathological Society (2020–2023), and currently as Vice President of the Biotic Stress Management Society (2024–present).

With over 50 publications in high-impact international journals, and as the author of the widely acclaimed textbook *Plant Bacteriology* (Kalyani Publishers), Dr. Mondal has established himself as a thought leader in plant pathology. He also serves as a reviewer for prestigious journals, including *BMC Microbiology*, *Plant Disease*, *Frontiers in Microbiology*, *MPMI*, *Microbial Pathogenesis*, *Scientific Reports*, *PMPP*, and *Plant Pathology*. His expertise has been showcased globally through invited lead talks and guest lectures in international conferences across the Philippines, Vietnam, China, Thailand, and the USA.

1 Historical Milestones in *Xanthomonas oryzae* pv. *oryzae* (*Xoo*) Research

The bacterial pathogen *Xanthomonas oryzae* pv. *oryzae* (*Xoo*) is responsible for bacterial blight (BB) of rice, one of the most destructive diseases affecting this staple crop in Asia and parts of Africa. For over a century, scientists have studied *Xoo* from its earliest identification, through classical plant pathology and epidemiology, to molecular dissection of its virulence mechanisms and the modern genomic era. This historical narrative traces the key milestones, discoveries, and people who have shaped *Xoo* research globally, emphasizing how each advance contributed to our understanding of pathogen biology, host resistance, and disease management.

1.1 FIRST REPORTS AND FOUNDATIONAL WORK

The earliest documentation of bacterial blight (BB) in rice dates back to the late nineteenth and early twentieth centuries, when agronomists and plant pathologists in Asia observed distinctive water-soaked lesions, wilting, and premature leaf drying during wet seasons. These early reports, primarily from Japan, highlighted a disease of serious economic importance that threatened rice paddies, although the causative organism remained elusive.

Bokura (1911) provided one of the first formal descriptions of the disease, identifying a bacterial agent he termed *Bacillus oryzae*, based on microscopic morphology and the characteristic symptomatic pattern observed in rice leaves. Bokura's work emphasized phenotypic characterization, which was the primary tool available for early plant pathologists. Subsequently, Ishiyama (1922) reclassified the bacterium as *Pseudomonas oryzae* and provided the first detailed pathological description, documenting elongated water-soaked lesions along leaf margins and interveinal chlorosis. Ishiyama's meticulous recording of lesion progression, seasonal patterns, and plant response laid the groundwork for recognizing BB as a microbial disease, although the pathogen's precise identity remained unresolved (Table 1.1).

Later, Nakata (1927) proposed *Bacterium oryzae*, retaining the species epithet to denote rice-specific pathogenicity. This work underscored the growing awareness that BB was caused by a host-adapted bacterial pathogen, distinct from other leaf diseases of rice.

A major taxonomic milestone occurred when Dowson (1943) formally established the bacterium within the genus *Xanthomonas*, naming it *X. oryzae*. Subsequent studies refined the classification further. For instance, Reitsma and Schure (1950) described a disease termed "kresek" in Indonesia, which was later

DOI: 10.1201/9781003714590-1

TABLE 1.1
Early Descriptions and Foundational Techniques

Year	Milestone/Contribution	Reference
1922	First pathological description of bacterial leaf blight in rice.	Ishiyama (1922)
1973	Standard lesion-length screening method established for BB.	Kauffman *et al.* (1973)
1975	Classical inheritance studies on resistance in Wase Aikoku 3.	Ezuka *et al.* (1975)
1979	First confirmed occurrence of BB in Africa.	Buddenhagen *et al.* (1979)
1985	Foundational host–pathogen studies in the Philippines.	Veracruz (1985)
1999	Comprehensive review of BB epidemiology and management.	Gnanamanickam *et al.* (1999)

confirmed as BB caused by *X. oryzae* pv. *oryzae*. Young *et al.* (1978) temporarily reclassified the pathogen as *Xanthomonas campestris* pv. *oryzae*, but Swings *et al.* (1990) reinstated *X. oryzae* and distinguished two pathovars: pv. *oryzae* causing bacterial blight and pv. *oryzicola* causing bacterial leaf streak. These taxonomic refinements provided a robust framework for systematic studies on BB biology, epidemiology, and control.

Parallel to these taxonomic advancements, foundational studies in pathology and genetics significantly shaped the understanding of host–pathogen interactions. Kauffman *et al.* (1973) introduced the lesion-length method, a standardized leaf-clipping inoculation technique for evaluating phenotypic response to BB by rice cultivars. This method enabled quantitative comparisons of cultivar responses, facilitating breeding programs aimed at improving BB resistance. Classical inheritance studies, such as those by Ezuka *et al.* (1975) using the Japanese cultivar Wase Aikoku 3, revealed Mendelian segregation of resistance traits, demonstrating that single dominant genes could govern host resistance.

The global significance of BB was realized when Buddenhagen *et al.* (1979) documented the first outbreaks in Africa, highlighting the transcontinental dispersal of *Xoo* and emphasizing the need for international monitoring and coordinated management. Later, comprehensive reviews, notably by Gnanamanickam *et al.* (1999), synthesized knowledge on BB epidemiology, cultural practices, chemical control measures, and early breeding strategies, integrating classical plant pathology with emerging molecular approaches.

These historical advancements notably in the areas of taxonomic refinements, and methodological innovations established a strong foundation for modern research on BB, enabling subsequent studies on pathogen biology, virulence mechanisms, and host resistance breeding.

1.2 HOST RESISTANCE GENE DISCOVERY, MAPPING, AND INTROGRESSION

One of the most transformative milestones in rice bacterial blight research was the cloning and functional characterization of the *Xa21* resistance gene (Table 1.2). In 1995, researchers in Pamela Ronald's laboratory at the University of California, Davis—including Guoliang Wang and Wenyuan Song—successfully isolated *Xa21*, demonstrating that it encodes a receptor kinase–like protein conferring broad-spectrum resistance against multiple strains of *X. oryzae* pv. *oryzae* (*Xoo*) (Song *et al.*, 1995). This discovery provided the first molecular insight into how rice perceives bacterial pathogens, setting the stage for modern understanding of plant immunity.

Building on this breakthrough, transgenic rice lines carrying *Xa21* were developed, producing the first stable *Xa21*-transgenic line in 1996. Subsequent field trials in India, the Philippines, and China confirmed the durability and stability of *Xa21*-mediated resistance in diverse environments (Datta *et al.*, 2002; Zhai *et al.*, 2004).

The cloning of *Xa21* catalyzed the discovery of additional resistance genes, both dominant and recessive. For instance, *xa5*, characterized by Iyer and McCouch (2004), employs a mechanism distinct from receptor kinases, whereas xa13 confers resistance *via* a loss-of-susceptibility mechanism, preventing *Xoo* from hijacking host sugar transporters critical for pathogen growth (Sanchez *et al.*, 1999; Oliva *et al.*, 2019). Dominant genes such as *Xa1*, *Xa3*, *Xa4*, *Xa7*, and *Xa21* encode receptor-like kinases or nucleotide-binding leucine-rich repeat (NLR) proteins, capable of either recognizing pathogen effectors directly or sensing effector-mediated modifications of host proteins, thereby triggering immune responses (Song *et al.*, 1995; Tian *et al.*, 2014).

The systematic incorporation of resistance genes into elite cultivars began in the 1970s and 1980s, spearheaded by the International Rice Research Institute (IRRI). Breeders developed near-isogenic lines (NILs) carrying single resistance genes, which allowed precise evaluation of *Xoo* pathotypes and the durability of resistance. Among the earliest and most widely deployed genes, *Xa4* significantly reduced BB incidence in commercial cultivars. However, its extensive use later imposed selective pressure on pathogen populations, highlighting the importance of diversifying resistance sources.

The exploration of wild rice species further accelerated resistance discovery. Amante-Bordeos *et al.* (1992) successfully introgressed resistance from *Oryza minuta*, demonstrating the potential of wild germplasm to broaden the genetic base of cultivated rice. Molecular marker development further enhanced precision breeding. For example, Blair and McCouch (1997) developed microsatellite markers linked to the *xa5* locus, which were subsequently fine-mapped (Blair *et al.*, 2003), allowing breeders to incorporate resistance efficiently through marker-assisted selection (MAS).

Understanding pathogen avirulence genes also provided critical insight into the gene-for-gene interaction underlying host resistance. Bai *et al.* (2000) demonstrated that specific *Xoo* effectors determine the activation of host resistance genes, linking

TABLE 1.2
Discovery, Genetic Mapping, and Deployment of Resistance Genes

Year	Milestone/Contribution	Reference
1992	Introgression of BB resistance from *Oryza minuta*.	Amante-Bordeos *et al.* (1992)
1997	Molecular markers linked to *xa5*.	Blair and McCouch (1997); Blair *et al.*, (2003)
2000	Avirulence gene roles in aggressiveness established.	Bai *et al.* (2000)
2002	Novel resistance gene identified in Minghui63.	Chen *et al.* (2002)
2003	Fine mapping of *xa5*.	Blair *et al.* (2003)
2008	BB resistance gene from *O. nivara* mapped.	Cheema *et al.* (2008)
2011	Recessive gene *xa34(t)* characterized.	Chen *et al.* (2011)
2016	Novel resistance gene from XM14 reported.	Busungu *et al.* (2016)
2016a, 2016b	Deployment of *Xa38* in Basmati through MAS.	Ellur *et al.* (2016a, 2016b)
2020	Discovery of *Xa46(t)*.	Chen *et al.* (2020)
2021	Executor gene *Xa7* identified.	Chen *et al.* (2021)
2024	*xa49(t)* discovered from *O. rufipogon*.	Bhatia *et al.* (2024)

molecular pathogenesis with practical breeding strategies. Subsequent research identified numerous novel resistance loci from diverse sources, including *Xa33* from *O. nivara* (Cheema *et al.*, 2008), *xa34(t)* (Chen *et al.*, 2011), and XM14 (Busungu *et al.*, 2016).

The introgression of *Xa38* into elite Basmati lines *via* MAS exemplifies the practical application of combining wild alleles with elite germplasm to enhance BB resistance (Ellur *et al.*, 2016a, 2016b). More recent breakthroughs include the identification of *Xa46(t)* (Chen *et al.*, 2020), the executor gene *Xa7* (Chen *et al.*, 2021), and *xa49(t)* from *O. rufipogon* (Bhatia *et al.*, 2024), highlighting the dynamic nature of host–pathogen co-evolution. Fine-mapping, functional characterization, and genome-editing tools have further accelerated the translation of these discoveries into breeding programs (Kumar *et al.*, 2012; Li *et al.*, 2016; Rashmi *et al.*, 2024).

Collectively, these studies demonstrate the progressive integration of classical genetics, molecular biology, and modern breeding strategies. From the first identification of race-specific genes to the targeted introgression of wild alleles and functional validation *via* genome editing, BB resistance research has enabled the development of rice cultivars capable of withstanding evolving *Xoo* populations, securing durable protection against one of the most destructive bacterial diseases in rice production.

1.2.1 Wild Rice as a Genetic Resource

Wild relatives of rice have contributed significantly to resistance breeding. Researchers such as Adhikari, Mew, and Teng identified durable resistance loci in *Oryza longistaminata* and *O. glaberrima*, generating new alleles suitable for pyramiding into elite cultivars. These findings underscore the importance of conserving global rice germplasm as a reservoir for resistance diversification, especially as *Xoo* populations continue to evolve under intensified agricultural practices.

1.3 *XOO* GENOMICS AND EFFECTOR BIOLOGY

Genomic approaches have revolutionized the understanding of *Xoo* biology, revealing how this vascular pathogen evolves, adapts, and executes virulence in rice ecosystems worldwide. Earlier work relied largely on classical microbiology, phenotypic characterization, and limited molecular markers. However, the availability of high-throughput, long-read, and integrative multi-omics technologies over the past decade has drastically expanded the scope of inquiry, uncovering detailed patterns of genome organization, effector diversification, population structure, and host–pathogen interactions. These advances have not only deepened fundamental knowledge of pathogen evolution but also accelerated resistance gene discovery, genome editing strategies, and epidemiological surveillance frameworks (Table 1.3).

A milestone in *Xoo* research occurred in 2008, when Salzberg *et al.* (2008) published the full genome sequence of strain PXO99A, revealing a single circular chromosome of 5,240,075 base pairs with numerous IS elements, multiple TAL effector genes, and major chromosomal rearrangements relative to other strains. This work demonstrated the plasticity of the *Xoo* genome, providing mechanistic insight into race diversification.

Seo *et al.* (2008) used a two-genome microarray to compare *Xoo* and *X. oryzae* pv. *oryzicola*, revealing differential regulation of *hrp* genes between the two pathovars, deepening understanding of secretion systems and regulatory circuits. High-throughput RNA sequencing revealed global transcriptional changes in rice upon *Xoo* infection, with key defense genes induced by *Xa21*- and *Xa23*-mediated resistance (Wang *et al.*, 2018). Proteomic profiling mapped changes in xylem sap proteins, highlighting defense signaling and cell wall modifications as key resistance mechanisms Tariq *et al.*, 2019).

Concurrently, comparative genomics across Asia and Africa has revealed patterns of both conservation and diversification. Chen *et al.* (2024) applied integrative transcriptomic and metabolomic approaches to rice cultivars challenged with diverse *Xoo* strains and documented that host responses differ significantly depending on strain-specific effector suites. Their work showed molecular signatures of susceptibility, particularly the induction of SWEET sugar transporters and of resistance pathways involving WRKY transcription factors, MAP kinases, ROS production, and cell wall reinforcement. These insights have demonstrated that even modest sequence variation among TAL effectors or host promoters can produce distinct interaction outcomes, thereby shaping the regional distribution of *Xoo* races and pathotypes.

TABLE 1.3
Major Milestones in Genomics, Effector Biology of *Xoo*

Year	Milestone/Contribution	Reference
2004	T3SS-effectors conceptualized as "double agents."	Alfano and Collmer (2004)
2005	Roles of beneficial bacteria in suppression of disease systems.	Compant *et al.* (2005)
2006	TAL-mediated activation of susceptibility via *xa13*.	Chu *et al.* (2006)
2006	HrcU determines substrate switching in T3SS.	Büttner *et al.* (2006)
2008	Diversity of non-TAL effectors among Korean strains.	Cho *et al.* (2008)
2009	Detailed review of T3SS structure and mechanism.	Büttner and He (2009)
2010	Biology, structure, and evolution of TAL effectors.	Boch and Bonas (2010)
2010	TAL effector–SWEET interactions clarified.	Chen *et al.* (2010)
2010	TAL effector DNA-binding code established.	Bogdanove *et al.* (2010)
2011	Whole-genome sequencing of *Xoo* provides evolutionary insights.	Bogdanove *et al.* (2011)
2012	XopR suppresses pattern-triggered immunity.	Akimoto-Tomiyama *et al.* (2012)
2013	XopN targets OsVOZ2 to dampen immunity.	Cheong *et al.* (2013)
2017	Editing *SWEET14* promoter to disrupt TAL recognition.	Blanvillain-Baufumé *et al.* (2017)
2018	Functional role of Xop effectors in Indian strains.	Verma *et al.* (2018)
2018	SWEET-mediated sugar fluxes in pathogen nutrition.	Bezrutczyk *et al.* (2018)
2019	XopQ–14-3-3 interactions; immune activation via phospho-mutants.	Deb *et al.* (2019)
2019	Review on suppression of immunity and virulence determinants.	Das *et al.* (2019)
2023	Large-scale host transcription reprogramming by Indian race 4 T3SS.	Amrutha Lakshmi *et al.* (2023)
2023	CRISPR-based promoter editing of *OsSWEET11*.	Bhagya Sree *et al.* (2023)
2024	Strain-specific transcriptome–metabolome profiling.	Chen *et al.* (2024)
2024	Environmental persistence niches of *Xoo* identified.	Duo *et al.* (2024)
2025	Long-read TALome sequencing of Pakistani isolates.	Ejaz *et al.* (2025)
2025	Expanded Xop effector functional roles in Indian *Xoo* populations.	Mondal *et al.* (2016)
2025	Functional validation of *OsRPM1* via VIGS.	Amrutha Lakshmi *et al.* (2025)

Environmental and ecological surveys have complemented these genomic insights. Duo *et al.* (2024) analyzed *Xoo* persistence in natural and agricultural reservoirs, identifying specific ecological niches, such as irrigation canals, weed hosts, and rice stubble, that favor long-term survival. Their work incorporated genomic tracking to link environmental isolates with field epidemics, revealing that certain lineages persist across seasons due to enhanced stress tolerance, biofilm formation, and niche-specific adaptation. This ecological-genomic integration supports epidemiological models emphasizing the influence of climate, cropping patterns, and landscape genetics on *Xoo* population turnovers.

Comparative genomic analyses across Indian, African, and Southeast Asian strains (Mondal *et al.*, 2020c; Poulin *et al.*, 2014; Timilsina *et al.*, 2020) have further revealed a dual evolutionary pattern: (i) a conserved set of core effectors required for fundamental virulence processes and (ii) variable genomic islands contributing to local adaptation, host specialization, and differential aggressiveness. These islands often harbor accessory TAL effectors, horizontally transferred genes, unique secretion system regulators, and mobile elements. Genome comparisons among Indian isolates such as IXO704 and IXOBB0003, Japanese strain MAFF311018, Korean strain KACC10331, and African strain BAI3 emphasized that the extent of genome plasticity strongly correlates with geographic context. Notably, African strains tend to harbor distinct TAL repertoires and insertion sequence compositions relative to Asian lineages, suggesting independent evolutionary trajectories shaped by host diversity and agroclimatic differences.

1.3.1 Effector Biology and Host Targets

Xoo virulence relies on T3SS, which delivers effectors into host cells. TAL effectors (e.g., PthXo1, PthXo3) bind specific promoter sequences in rice to induce SWEET sugar transporter genes, promoting pathogen nutrition and disease progression (Moscou and Bogdanove, 2009; Streubel *et al.*, 2013; Mondal *et al.*, 2021).

Non-TAL effectors, including XopR, XopF, and XopL, manipulate host cellular processes, suppressing reactive oxygen species (ROS) and altering ubiquitination pathways (Mondal *et al.*, 2016; Song and Yang, 2010; Pruneda *et al.*, 2016). Together, these effectors orchestrate a molecular "double agent" strategy, simultaneously subverting host immunity and enhancing pathogen virulence (Alfano and Collmer, 2004).

Disruption of TAL effector binding sites in SWEET promoters has been employed to generate resistant rice lines, demonstrating the functional relevance of effector–host interactions in practical disease management (Blanvillain-Baufumé *et al.*, 2017).

Additionally, studies integrating TAL effector repertoires with rice genomic targets generated predictive models for effector binding, informing both resistance gene discovery and genome editing strategies (Grau *et al.*, 2019).

1.3.2 Mobile Genetic Elements and Pathogenicity Islands

Mobile genetic elements are key drivers of genome evolution and virulence in plant pathogenic bacteria. These include insertion sequences (IS elements), transposons, plasmids, prophages, and pathogenicity islands. They promote horizontal gene transfer, gene duplication, and genome rearrangements, allowing pathogens to adapt rapidly to host resistance.

IS elements are small DNA sequences that can move within the genome. They often occur near transcription activator-like (TAL) effector gene clusters and promote duplications and rearrangements. This leads to variation in effector genes, which helps pathogens overcome host resistance by altering recognition (Boch and Bonas, 2010; Salzberg et al., 2008). Transposons, which are larger elements, can carry additional genes, including virulence factors, and help in their movement across the genome.

Plasmids are extra-chromosomal DNA molecules that frequently carry accessory virulence genes, such as those encoding toxins and secretion systems. They enable rapid spread of pathogenicity traits between strains and species, contributing to the emergence of new pathotypes (Vivian et al., 2001). Prophages, which are bacteriophage genomes integrated into bacterial chromosomes, also contribute to virulence by carrying fitness and regulatory genes and enabling gene transfer through transduction.

Pathogenicity islands are large genomic regions acquired through horizontal gene transfer. They contain clusters of virulence genes, including those for the type III secretion system and effector proteins. Pathogenicity islands are often associated with mobility genes, indicating their ability to move within or between genomes (Hacker and Kaper, 2000). Gain or loss of such gene can result in major changes in virulence and host range.

1.3.3 Fundamentals of TAL Effectors in *Xoo*

The landmark genome sequencing of *Xoo* by Bogdanove *et al.* (2011) provided the first comprehensive framework for understanding genome organization, particularly the architecture of type III secretion system (T3SS) *hrp/hrc* clusters, families of transcription activator-like effectors (TAL effectors), and mobile elements underlying genome plasticity. Their work clarified the modular nature of TAL effectors, each consisting of repeat arrays with repeat-variable diresidues (RVDs) dictating DNA-binding specificity, thus laying the foundation for subsequent computational prediction tools and gene-editing strategies. This sequence also revealed the prevalence of insertion sequence elements and the numerous genes encoding hypothetical proteins, suggesting that only a fraction of pathogenicity-related factors had been functionally characterized at that time. Pioneering studies in the early 2010s characterized TAL effectors at a mechanistic level. Boch *et al.* (2009) revealed that TAL effectors bind promoter sequences in host susceptibility genes with remarkable specificity determined by their RVD codes. This breakthrough established the effector–host gene paradigm and enabled the rational design of resistance strategies targeting effector-binding elements (EBE) sites. Parallel work by Moscou and Bogdanove (2009) led to computational models predicting TAL effector binding sites, accelerating the identification of host targets and enabling genome-editing strategies that modify

EBEs without disrupting endogenous gene function. These efforts collectively transformed the field, enabling precision manipulation of host–pathogen interactions. The advent of long-read sequencing technologies, combined with improved assembly algorithms, has significantly enhanced genome completeness and resolution. Ejaz *et al.* (2025) generated high-quality assemblies of two *Xoo* isolates from the basmati-growing region of Pakistan and demonstrated that each genome (~5.12 Mb) contained 18 TAL effectors, including three predicted pseudo-TALs. These long-read assemblies clarified complex genomic regions containing tandem repeats, duplicated effector loci, and structural variants that were previously obscured by short-read technologies. In a surprising discovery, Ejaz *et al.* (2025) reported that TAL profiles were nearly identical among isolates separated geographically, suggesting strong selective pressures maintaining conserved effector sets within particular rice-growing ecosystems (Table 1.3). Their study implied that in certain ecological niches, host genotypes and agricultural landscapes enforce stabilizing selection rather than diversifying evolution on TAL repertoires.

These genetic discoveries underscore the synergistic value of combining promoter editing, gene pyramiding, and genomic surveillance to maintain long-term resistance under dynamic pathogenic pressures.

1.3.4 Multi-Omics Insights into Host Response and Metabolic Reprogramming

Recent studies have leveraged transcriptomics, proteomics, and metabolomics to unravel the complexity of rice immune responses during *Xoo* infection. Chen *et al.* (2024) used a multi-omics framework to analyze the cultivar CBB23 carrying the broad-spectrum resistance gene *Xa23* when challenged with strains AH28 and PXO99A. Their transcriptome data revealed more than 8,000 differentially expressed genes, including key transcription factor families such as WRKY, NAC, and AP2/ERF that orchestrate early defense signaling. Enhanced expression of kinases involved in MAPK signaling, calcium-dependent pathways, and redox-balancing networks further demonstrated the activation of robust early immune signaling.

Proteomic analyses revealed alterations in xylem sap composition, including upregulation of pathogenesis-related proteins, peroxidases, and cell wall–modifying enzymes such as expansins and pectin methylesterases. These proteins collectively strengthen vascular defenses and inhibit bacterial movement through the xylem.

Metabolomic profiling showed increases in terpenoids, flavonoids, and defensive amino acid derivatives, indicating extensive metabolic reprogramming that complements immune signaling. Critically, multi-omics convergence demonstrated that SWEET transporter induction—central to susceptibility in many rice varieties—is effectively suppressed in *Xa23*-bearing plants, providing a mechanistic basis for durable resistance.

These multi-omics frameworks offer a global understanding of host–pathogen molecular dynamics and serve as blueprints for precision breeding and targeted manipulation of resistance pathways.

1.3.5 Genome-Wide Association and QTL Mapping

In 2012, a genome-wide association study (GWAS) involving 300 indica accessions was conducted at the China National Rice Research Institute (Zhang *et al.*, 2012). Their work identified several novel QTLs, including XA31 and XA32, conferring partial but stable resistance against multiple *Xoo* strains. These loci were subsequently validated in additional populations and now represent promising candidates for pyramiding.

In 2013, bi-parental mapping populations were used to identify *Xa* genes effective against Indian pathotypes (Thakur *et al.*, 2013). Their research uncovered allelic variants of *Xa27* that displayed enhanced field resistance under North Indian agro-climatic conditions. These mapping efforts demonstrated that localized breeding strategies must account for region-specific *Xoo* diversity, an insight that has become increasingly relevant, given the rapid evolution of pathogen populations.

Together, advances in genomics, long-read sequencing, functional analyses, and multi-omics have profoundly reshaped the understanding of *Xoo* evolution, virulence, and host interactions. Integrating these datasets enables more accurate prediction of pathogen dynamics, guides durable resistance design, and strengthens epidemiological preparedness. As genome editing tools and sequencing technologies continue to advance, the future of *Xoo* research lies in predictive disease modeling, landscape-level genomic surveillance, and synthetic engineering of resistance traits that remain robust against pathogen evolution.

1.4 PATHOGEN DIVERSITY AND EVOLUTION

1.4.1 Emergence of Race Concepts

By the 1950s and 1960s, plant pathologists conducting extensive disease surveys observed variable severity of BB among different rice cultivars, even within the same field. Such variability indicated that *Xoo* populations were not homogeneous. Researchers in Japan, the Philippines, and India documented these differences and proposed the concept of "races" or pathotypes, defined by the differential response of rice varieties. These early pathotype designations became critical for resistance breeding, enabling the selection of varieties effective against local pathogen populations.

1.4.2 Genomic Plasticity

The genetic diversity of *Xoo* is largely shaped by horizontal gene transfer, insertion sequences (IS elements), and integrons, which allow rapid acquisition of virulence and antibiotic resistance traits (Mazel, 2006; Rowe-Magnus and Mazel, 2002; Stalder *et al.*, 2012). Comparative genomics of Indian and global strains highlighted the co-evolutionary arms race between pathogen effectors and host resistance genes (Mondal *et al.*, 2020c; Poulin *et al.*, 2014; Timilsina *et al.*, 2020).

Emergence of new race capable of overcoming resistance genes like improved Sambha Mahsuri underscores the need for dynamic disease management strategies (Tiwari *et al.*, 2025). Genomic analyses of virulent strains, such as PXO99A, revealed numerous IS elements, multiple TAL effector genes, and chromosomal rearrangements, collectively explaining the rapid diversification of *Xoo* races (Lee *et al.*, 2005; Salzberg *et al.*, 2008; Poulin *et al.*, 2014; Midha *et al.*, 2017).

1.4.3 Mobile Genetic Elements and Evolutionary Dynamics

Long-read sequencing has also facilitated detailed mapping of insertion sequences (IS elements), prophages, and other mobile elements that influence genome evolution. Ejaz *et al.* (2025) demonstrated that mobile genetic elements shape TAL effector diversification, mediate gene rearrangements, and reflect historical bacteriophage interactions. The presence of prophage sequences in multiple strains indicates past phage-mediated horizontal gene transfers that may have introduced new regulatory elements or effector domains. IS element–driven rearrangements also contribute to structural variation in effector loci, offering a potential mechanism for rapid adaptation in response to host resistance.

Understanding these mobile elements is critical because they underpin the pathogen's capacity to evolve under natural selection, pesticide pressure, and resistance gene deployment. Characterizing these patterns allows more accurate forecasting of virulence shifts, guiding resistance breeding and crop rotation planning.

1.5 RAPID RACE MONITORING AND SURVEILLANCE

The rapid evolution of *Xoo* requires proactive monitoring of pathogen populations. Researchers now employ promoter-edited tracer lines to detect dominant TAL effectors in field isolates, providing real-time information on prevalent pathotypes. Liu *et al.* (2024) demonstrated the effectiveness of this approach in northern India, where the distribution of TAL effectors such as PthXo1, PthXo2, and PthXo3 was mapped across multiple districts. This knowledge guided the strategic deployment of promoter-edited lines and pyramided resistance genes.

High-throughput sequencing and metagenomic analyses complement surveillance efforts by revealing accessory virulence plasmids, prophage elements, and rare TAL effector variants. Data from these approaches feed into predictive models that integrate pathogen genomics, environmental factors, and resistance gene deployment, enabling dynamic adjustment of breeding strategies and field-level management.

1.5.1 Molecular Tools for Disease Surveillance

Molecular detection techniques, including polymerase chain reaction (PCR)–based diagnostics, RFLP markers, and whole-genome sequencing, have improved early detection and tracking of *Xoo* pathotypes (Lu *et al.*, 2014; Midha *et al.*, 2017; Liu *et al.*, 2021). Additionally, AI-based image analysis facilitates rapid field-level detection of BB symptoms, aiding precision agriculture and timely intervention (Mondal and Amrutha Lakshmi, 2020).

1.5.2 Integration of AI, High-Throughput Phenotyping, and Genotyping

Artificial intelligence and machine learning are increasingly employed to monitor *Xoo* disease dynamics and resistance efficacy. High-throughput phenotyping platforms utilize drones, robotics, and imaging sensors to capture disease progression in large experimental and commercial fields. AI algorithms analyze leaf images, xylem sap indicators, and growth metrics to quantify infection severity and host response.

When coupled with high-resolution genotyping, these technologies enable the construction of genotype–phenotype maps linking TAL effector repertoires, promoter edits, and resistance loci to observed field performance. Predictive models integrating pathogen genomics, host genetics, and environmental data allow researchers to forecast pathotype emergence and optimize the deployment of promoter-edited or pyramided lines in targeted regions (Mondal *et al.*, 2014, 2023; Qi *et al.*, 2025).

1.5.3 Seed Transmission and Epidemiological Studies

Epidemiological studies in the 1960s and 1970s revealed that contaminated seed acted as a primary reservoir for long-distance dissemination of *Xoo*. Irrigation water, field tools, and cultural practices were identified as factors facilitating epidemic spread. These insights shaped early BB management recommendations, emphasizing the use of clean seed, field sanitation, crop rotation, and vigilant monitoring to limit pathogen transmission.

1.6 DIAGNOSTICS AND DISEASE MANAGEMENT STRATEGIES

1.6.1 Molecular Diagnostics

Real-time PCR, loop-mediated isothermal amplification (LAMP), and portable sequencing platforms enabled rapid, field-level detection of *Xoo*.

AI-based image recognition systems were deployed to detect early bacterial leaf blight (BLB) symptoms from leaf images in experimental field trials, demonstrating potential for precision disease monitoring (Kirti *et al.*, 2022).

Recent studies detected *Xoo* in irrigation water, rice stubble, and non-host grasses, emphasizing the ecological persistence of the pathogen and informing quarantine strategies (Duo *et al.*, 2024).

1.6.2 Biological Control and Plant Growth Promotion

Beneficial rhizobacteria, including *Bacillus* and *Pseudomonas* species, have been employed to induce systemic resistance and enhance crop growth (Kloepper *et al.*, 2004; Pieterse *et al.*, 2014; Mondal *et al.*, 2010). Production of antimicrobial metabolites such as lipopeptides and 2,4-diacetylphloroglucinol contributes to the suppression of *Xoo* populations in the rhizosphere (Ongena and Jacques, 2008; Velusamy *et al.*, 2006). Recent studies also explore endophytic bacterial consortia and mycosynthesized nanoparticles as novel biocontrol agents against BB (Resti *et al.*, 2020; Shobha *et al.*, 2020; Su *et al.*, 2023).

1.6.3 Chemical Control

Although chemical agents, including streptomycin and copper-based compounds, have been historically used to manage BB, concerns regarding resistance development and environmental safety have limited their application (McManus and Jones, 1994; Singh *et al.*, 1980; Sundin and Bender, 1993). Integrated approaches combining genetics, biocontrol, and cultural practices are now preferred (Nasir *et al.*, 2019; Shekhar *et al.*, 2020).

1.6.4 Genome Editing and Advanced Resistance Strategies

The discovery that many TAL effectors activate specific host susceptibility genes (*S* genes), particularly *OsSWEET* family members, has transformed resistance breeding strategies. Promoter editing of SWEET genes represents one of the most impactful interventions in recent years. Liu *et al.* (2024) used CRISPR/Cas9 to edit promoter regions of *OsSWEET11a*, *OsSWEET13*, and *OsSWEET14* in multiple rice backgrounds and identified the TAL effectors responsible for virulence across fifty *Xoo* isolates. Their promoter-edited lines displayed broad-spectrum resistance, confirming that targeted disruption of TAL-binding elements (EBEs) can effectively block pathogen exploitation of host sugar transport.

Similarly, in the ASD16 variety, *OsSWEET11* was successfully edited by Bhagya Sree *et al.* (2023), producing homozygous lines resistant to strains expressing the major TAL effector PthXo1. These studies collectively highlight the durability and precision of promoter editing relative to classical resistance (*Xa*) genes, which often confer race-specific resistance vulnerable to TAL diversification.

Han *et al.* (2025) expanded the concept by rationally designing multi-element promoter modifications to confer resistance against multiple pathogens simultaneously, while minimizing fitness penalties. Their work demonstrates that synthetic engineering of promoter architecture—through multiplex EBE disruption—can create robust yet physiologically stable resistance. Substantial advancements in Arabidopsis and rice further indicated that even subtle modifications within cis-regulatory modules can profoundly alter pathogen response.

The recent identification of *Xa50(t)* emphasizes the need for continuous exploration of natural resistance loci. *Xa50(t)* activates defense signaling distinct from classical *Xa* pathways, induces rapid ROS formation, and strengthens vascular barriers. When combined with other *Xa* genes, *Xa50(t)* contributes additively to durable and field-relevant resistance. The integration of genome editing and molecular breeding offers new avenues to control BB. CRISPR/Cas-mediated promoter editing of *OsSWEET11* successfully confers resistance by disrupting TAL effector recognition sites (Bhagya Sree *et al.*, 2023). Functional validation of host immune regulators, such as *OsRPM1*, via virus-induced gene silencing (VIGS) demonstrates the potential for exploiting host immunity for BB control (Amrutha Lakshmi *et al.*, 2025). These strategies represent the frontier of rational resistance engineering, combining precise genome editing with conventional breeding and effector-informed molecular knowledge. Multiple strains exhibited CRISPR-Cas loci, with strain-specific spacer sequences. These patterns reflected historical phage infection events and provided insights into the pathogen's evolutionary history (Zhang *et al.*, 2014).

1.6.5 Synthetic Biology and Engineering of Resistance

Given the limitations of natural resistance genes, synthetic biology offers new avenues for designing multi-layered defense systems against *Xoo*. By combining promoter-edited susceptibility loci, newly identified resistance genes such as *Xa50(t)*, and synthetic regulatory modules, researchers have developed rice lines with robust defense responses. These lines integrate edited effector-binding elements in *SWEET* gene promoters, stacked resistance genes, and transcriptional regulators that modulate immune signaling to prevent excessive fitness costs.

Experimental evaluation of these synthetic constructs shows rapid pathogen recognition and activation of defense pathways, including hypersensitive response, production of antimicrobial compounds, and reinforcement of vascular barriers. Synthetic biology approaches also enable predictive modeling of pathogen evolution, allowing breeders to anticipate virulence strategies and design rice lines capable of withstanding emergent pathotypes. These innovations mark a significant shift from traditional breeding to rational engineering of durable resistance.

1.6.6 Toward Climate-Resilient Disease Management

Climate change is altering the dynamics of *Xoo* infection and rice susceptibility. Increased temperatures, erratic rainfall, and extreme weather events are projected to exacerbate bacterial blight epidemics in vulnerable regions. Contemporary research integrates climate modeling with genomic and phenotypic analyses of both host and pathogen to develop climate-resilient management strategies.

Field trials conducted in heat- and drought-prone areas with promoter-edited lines, pyramided resistance genes, and synthetic constructs indicate that engineered lines maintain resistance under elevated temperature and water stress conditions. Lesion lengths and bacterial population densities were significantly lower compared to conventional cultivars, highlighting the potential for combining genetic resistance with environmental adaptation to sustain productivity under changing climatic conditions.

1.6.7 Gene Pyramiding toward Durable Resistance

Pyramiding multiple resistance genes through marker-assisted selection remains a cornerstone strategy for durable BB resistance (Singh *et al.*, 2016; Sagar *et al.*, 2018, 2020). Modern approaches include CRISPR/Cas9-mediated genome editing to engineer broad-spectrum resistance by targeting susceptibility genes or stacking multiple resistance loci (Li *et al.*, 2016; Oliva *et al.*, 2019; Ramasamy *et al.*, 2025).

The deployment of pyramided BB resistance genes had become a central strategy in global rice improvement programs. At the International Rice Research Institute (IRRI), combinations such as *Xa4* + *xa5* + *Xa21* were introgressed into elite backgrounds, producing lines that exhibited broad-spectrum and durable resistance under multi-season field evaluations. Extensive testing across diverse agro-ecosystems in the Philippines and Vietnam demonstrated that these pyramids significantly delayed pathogen adaptation, maintaining effectiveness against local pathotypes with high virulence diversity (Sanchez *et al.*, 2000).

Parallel advancements occurred in India, where Indian Council of Agricultural Research (ICAR) breeding groups successfully combined *xa5* + *xa13* + *Xa21* into high-yielding varieties, including PR121, enabling robust resistance against the rapidly evolving North Indian pathotypes of *Xoo* (Tiwari *et al.*, 2015). These pyramids capitalized on complementary resistance mechanisms: *xa5* reduced susceptibility by modulating transcriptional elongation, *xa13* hindered pathogen exploitation of the SWEET-mediated xylem sugar flux, and *Xa21* triggered strong pattern-recognition-mediated immunity. Their combined effect provided durable defense while maintaining agronomic performance in intensive rice–wheat zones.

A major milestone in BB-resistant Basmati improvement came through the long-term breeding effort at different institutes of ICAR including ICAR-NRRI, ICAR-IIRR and ICAR-IARI. The Pusa Basmati improvement program undertaken at ICAR-IARI to integrate durable, multi-gene BB resistance without compromising hallmark grain and aroma traits. This initiative led to the release of Pusa Basmati 1718 (PB1718), carrying the pyramided resistance genes *xa13* + *Xa21*, offering strong field protection against major BB pathotypes. Building on this foundation, ICAR-IARI developed next-generation pyramided Basmati derivatives through marker-assisted backcross breeding coupled with stringent grain-quality recovery.

Two of the most impactful recent releases are Pusa Basmati 1847, an improved derivative of PB1509, and Pusa Basmati 1885, derived from PB1121. Both lines carry *xa13* + *Xa21*, and in select advanced versions, additional BB and blast resistance loci were introduced depending on the target genetic background. Multi-location evaluations across Haryana, Punjab, Delhi, Uttar Pradesh, and select zones of Uttarakhand demonstrated consistent suppression of lesion expansion, reduced bacterial loads, and significant yield stability under disease pressure (Singh *et al.*, 2016). Simple sequence repeat (SSR), single nucleotide polymorphism (SNP), and functional markers accelerated pyramiding programs. Marker-assisted selection (MAS) allowed breeders to track multiple *Xa* genes simultaneously, shortening breeding cycles from 7–8 years to 3–4 years (Singh *et al.*, 2016).

These cultivars exemplify how precision pyramiding, when integrated with high-resolution genotyping and background selection, can produce commercially viable Basmati varieties that combine exporters' quality standards with strong and durable BB resistance. The durability of these pyramids also reflects a strategic match between host genetics and regional pathogen ecology: North-Western India is dominated by IXO-series isolates with conserved TAL effector repertoires, for which *xa13-Xa21* combinations remain highly effective. As a result, Pusa Basmati 1847 and 1885 have rapidly gained adoption among farmers, who benefit from improved disease resistance without compromising market value, aroma, or milling quality.

The continuous refinement of pyramiding strategies at IRRI, ICAR, and state agricultural universities (SAUs) underscores the importance of integrating genomic surveillance of *Xoo* populations with breeding interventions. This approach ensures that resistance genes deployed in elite cultivars remain aligned with pathogen evolution, thereby enhancing long-term management of bacterial blight in both basmati and non-basmati production systems.

1.6.8 Translational Breeding and Deployment Strategies

The ultimate objective of research on *Xoo* is to ensure that advances in molecular biology, genomics, and plant–microbe interactions are translated into tangible, field-level solutions for the sustainable management of bacterial blight (BB) of rice. Translational breeding represents a critical bridge between laboratory discoveries and farmer adoption, integrating molecular innovations with conventional breeding wisdom, agronomic performance, and socio-economic realities. In recent years, rice improvement programs have increasingly shifted from single gene resistance approaches toward multi-layered, knowledge-driven strategies that combine promoter-edited susceptibility (*S*) genes, classical resistance (*R*) genes, and engineered regulatory modules to achieve durable and broad-spectrum resistance.

Modern breeding programs now recognize that resistance durability is rarely achieved through the deployment of a single resistance source. Instead, the focus has moved toward pyramiding multiple resistance mechanisms that act at different stages of pathogen infection and colonization. Promoter editing of SWEET genes, which prevents transcription activator-like effector (TALE)–mediated induction of host susceptibility, has emerged as a powerful strategy for blocking pathogen nutrition without imposing major fitness costs on the host. When such edited S genes are combined with well-characterized *R* genes such as *Xa21* and *Xa50(t)*, the resulting cultivars possess both basal and inducible defense layers. This stacking of resistance mechanisms reduces the likelihood that *Xoo* populations can rapidly adapt, as the pathogen must overcome multiple genetic barriers simultaneously.

Classical resistance genes continue to play a foundational role in translational breeding for BB resistance. Genes such as *Xa21*, which encodes a pattern recognition receptor recognizing conserved microbial signatures, have demonstrated broad spectrum and relatively durable resistance across diverse environments. Similarly, newly identified or underutilized genes such as *Xa50(t)* contribute additional layers of defense that complement both edited S genes and other *R* genes. Translational breeding strategies emphasize not only the identification of such genes but also their precise introgression into elite genetic backgrounds through marker-assisted selection (MAS) and genomic selection. This ensures that resistance traits are incorporated without compromising yield, grain quality, or adaptation to local agroecologies.

Synthetic regulatory modules represent an emerging frontier in translational breeding. These modules are designed to fine-tune host defense responses by modulating gene expression spatially and temporally, thereby minimizing negative trade-offs between resistance and growth. By engineering promoters that are activated only upon pathogen attack or under specific environmental conditions, breeders can achieve responsive resistance that is both efficient and resource-conserving. When integrated with genome-edited *S* genes and classical *R* genes, such synthetic systems enable a more sophisticated and resilient defense architecture within the rice plant.

A defining feature of translational breeding is rigorous multi-location field evaluation. Resistance identified under controlled conditions must be validated across diverse agroclimatic zones to assess durability, stability, and agronomic performance. Multi-location trials provide insights into genotype × environment interactions, allowing breeders to identify varieties that maintain resistance and yield stability under varying temperature regimes, rainfall patterns, soil types, and pathogen pressures. These trials also help detect early signs of resistance breakdown, enabling timely refinement of breeding strategies before large-scale deployment.

Yield stability and farmer acceptability remain central considerations in the deployment of BB-resistant rice varieties. Translational breeding programs therefore integrate resistance evaluation with comprehensive assessments of yield components, stress tolerance, and grain quality attributes such as cooking quality, aroma, and market preference. Consumer-preferred traits are particularly important for ensuring adoption, as even highly resistant varieties may fail if they do not meet culinary or cultural expectations. The integration of resistance breeding with quality trait improvement reflects a holistic approach that aligns scientific innovation with real-world demand.

International and national collaborations play a pivotal role in accelerating translational breeding and deployment. Partnerships among institutions such as IRRI, ICAR, SAUs, and private seed companies facilitate the exchange of germplasm, knowledge, and technologies. Such collaborations enable the pyramiding of resistance genes from diverse genetic sources and ensure that breeding outputs are tested across multiple epidemiological contexts. Shared field evaluation networks enhance the robustness of resistance assessments and promote harmonization of breeding objectives across regions.

Germplasm exchange underpins the success of these collaborative efforts. Access to diverse landraces, wild relatives, and pre-breeding lines expands the resistance gene pool and reduces dependence on a narrow set of resistance sources. Translational breeding frameworks emphasize responsible germplasm sharing, supported by clear regulatory and intellectual property guidelines, to ensure equitable access and long-term sustainability. Through coordinated efforts, resistance genes identified in one region can be rapidly mobilized and adapted for deployment in other BB-prone rice-growing areas.

The integration of genomic surveillance into translational breeding represents a transformative advancement in BB management. Continuous monitoring of *Xoo* population structure, virulence profiles, and effector diversity provides critical intelligence for breeding decision-making. By aligning resistance gene deployment with real-time data on pathogen evolution, breeders can anticipate potential resistance breakdowns and proactively introduce complementary genes or edited alleles. This predictive approach marks a shift from reactive breeding, where resistance is addressed only after widespread failure, to a forward-looking strategy that seeks to stay ahead of pathogen adaptation.

Multi-layered resistance strategies are increasingly complemented by climate adaptation considerations. Climate change influences the epidemiology of bacterial blight by altering temperature, humidity, and rainfall patterns, which in turn affect pathogen survival, dispersal, and aggressiveness. Translational breeding programs therefore integrate BB resistance with traits conferring tolerance to abiotic stresses such as heat, drought, and submergence. By developing varieties that are resilient to both biotic and abiotic stresses, breeders enhance the overall stability of rice production systems under changing climatic conditions.

Emerging technologies are accelerating the pace and precision of translational breeding. Genome editing tools enable rapid modification of target loci without the lengthy backcrossing cycles associated with conventional breeding. High-throughput phenotyping platforms, including imaging, remote sensing, and automated disease scoring, allow large populations to be evaluated efficiently and objectively. Integrative omics approaches combining genomics, transcriptomics, proteomics, and metabolomics provide a systems-level understanding of host–pathogen interactions and identify novel targets for resistance enhancement. Together, these technologies shorten breeding cycles and increase the probability of delivering robust, field-ready cultivars.

Despite these advances, the dynamic evolution of *Xoo* remains a persistent challenge. The pathogen's diverse and adaptable effector arsenal enables it to circumvent host defenses, particularly when resistance deployment lacks strategic coordination. Translational breeding must therefore be accompanied by sustainable management strategies, including resistance gene rotation, mosaic deployment across landscapes, and integration with agronomic practices that reduce disease pressure. Such strategies slow the selection of virulent pathogen strains and extend the effective lifespan of resistant cultivars.

Capacity building and knowledge transfer are essential components of successful deployment. Breeders, pathologists, extension personnel, and farmers must be equipped with the skills and information necessary to implement resistance strategies effectively. Training programs, participatory varietal selection, and on-farm demonstrations strengthen the link between breeding innovations and field-level impact. By engaging stakeholders across the rice value chain, translational breeding initiatives foster shared ownership and informed adoption of BB-resistant varieties.

Policy support and regulatory frameworks also influence the translation of research into practice. Clear guidelines for the evaluation and release of genome-edited crops, harmonized across regions, are crucial for timely deployment. Public communication and transparency regarding the safety and benefits of advanced breeding technologies help build consumer trust and acceptance. Translational breeding thus operates at the interface of science, policy, and society, requiring coordinated action beyond the laboratory and breeding station.

In conclusion, translational breeding and deployment strategies for bacterial blight resistance exemplify a paradigm shift toward proactive, predictive, and integrative crop improvement. By combining promoter-edited *S* genes, classical and novel *R* genes, synthetic regulatory modules, and genomic surveillance, modern breeding programs aim to deliver rice cultivars that are resilient, productive, and acceptable to farmers and consumers alike. Continuous monitoring of *Xoo* evolution, coupled with international collaboration and emerging technologies, will be essential to sustain these gains. As emphasized by recent studies (Qi *et al.*, 2025; Nganga *et al.*, 2025; Ramasamy *et al.*, 2025), safeguarding global rice production against bacterial blight requires not only scientific innovation but also strategic deployment and long-term stewardship of resistance resources.

2 Origin, Geographical Distribution, and Losses

Bacterial leaf blight (BB), caused by *Xanthomonas oryzae* pv. *oryzae* (*Xoo*), is recognized as one of the most serious bacterial diseases of rice worldwide (Mew, 1987, 1992; Mondal, 2011, 2017; Mondal *et al.*, 2022b). The disease was first observed in Kyushu, Japan, during 1884–1885, and its bacterial etiology was later conclusively established in 1922 (Ishiyama, 1922). Since its initial recognition, BB has emerged as a persistent and destructive constraint to rice production, particularly in regions characterized by warm temperatures, high humidity, and intensive rice cultivation. The importance of BB is further magnified by the staple role of rice in global food security, making the disease not merely an agronomic problem but also a socio-economic concern affecting farmer livelihoods and national food supplies.

The pathogen *Xoo* is highly adapted to rice and exhibits remarkable genetic diversity and evolutionary potential. This adaptability enables the bacterium to persist across diverse agroecological zones and to overcome host resistance when deployed without strategic planning. Consequently, understanding the geographical distribution and yield losses associated with BB is essential for designing effective management strategies and guiding breeding and policy interventions.

2.1 NATIONAL DISTRIBUTION

Bacterial blight is widely prevalent across most rice-growing regions of India and is regarded as one of the most consistently occurring diseases of the crop. Each year, the disease manifests at moderate to severe levels in nearly all major rice-producing states (Figure 2.1). These include Andhra Pradesh, Assam, Andaman and Nicobar Islands, Bihar, Haryana, Himachal Pradesh, Jharkhand, Karnataka, Kerala, Madhya Pradesh, Maharashtra, Manipur, Meghalaya, Mizoram, Nagaland, Odisha, Punjab, Rajasthan, Tamil Nadu, Tripura, Uttarakhand, Uttar Pradesh, and West Bengal. The widespread occurrence of BB across such a broad geographical range highlights the adaptability of the pathogen and the vulnerability of rice-based production systems.

The intensity and frequency of BB outbreaks within the country vary depending on climatic conditions, varietal susceptibility, cropping patterns, and management practices. States with high rainfall, prolonged periods of leaf wetness, and intensive mono-cropping of rice often experience more severe disease pressure. Irrigated lowland ecosystems, which dominate rice cultivation in many parts of India, provide favorable conditions for pathogen survival, multiplication, and dissemination. In contrast, upland and rainfed systems may experience sporadic outbreaks, often linked to specific weather events such as heavy monsoon rains or storms.

DOI: 10.1201/9781003714590-2

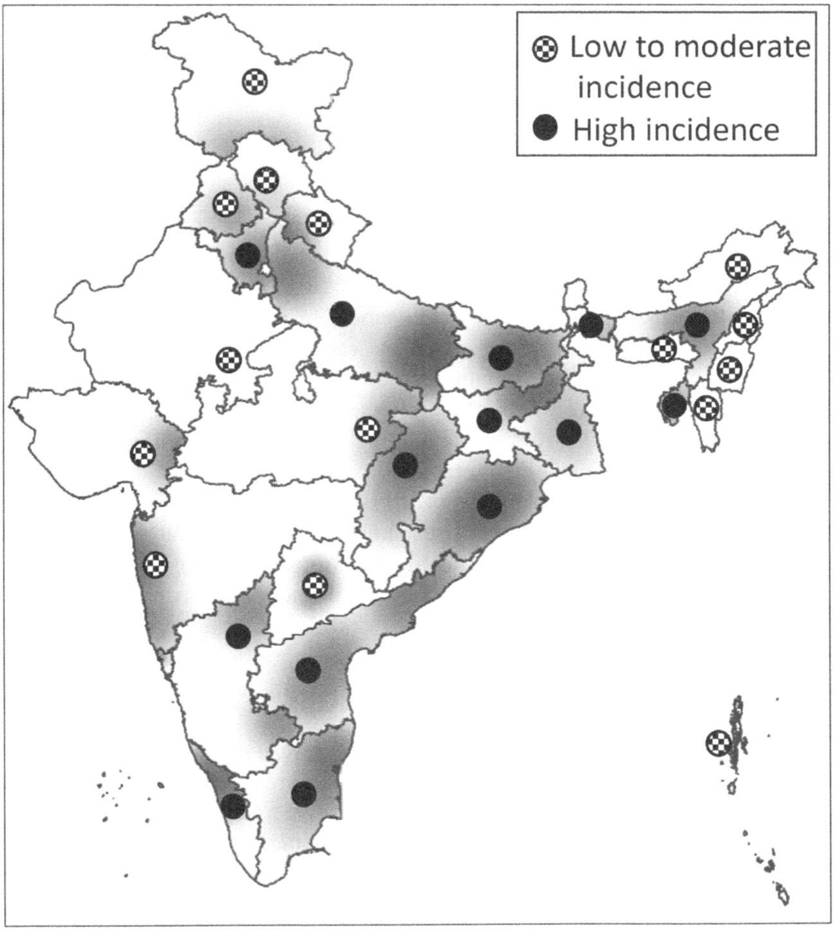

FIGURE 2.1 Distribution of BB in Indian states. Circle with dotted fill indicates low to moderate incidence while that of solid black fill indicates high incidence.

In eastern and northeastern India, including states such as West Bengal, Assam, Bihar, Jharkhand, Odisha, Manipur, and Tripura, BB is a recurring problem due to high humidity and the widespread cultivation of susceptible high-yielding varieties. Similarly, southern states such as Tamil Nadu, Andhra Pradesh, and Karnataka frequently report BB incidence, particularly during the wet season. Northern states, including Punjab, Haryana, and Uttar Pradesh, despite comparatively drier climates, also experience significant disease pressure under irrigated conditions, especially when favorable temperatures coincide with high nitrogen fertilization.

The persistent national-scale distribution of BB underscores the limitations of relying solely on chemical or cultural control measures. It also highlights the importance of deploying resistant varieties tailored to regional pathogen populations and agroclimatic conditions. Continuous monitoring of disease distribution at the national level is therefore critical for early warning, resistance breeding, and strategic deployment of management interventions.

2.2 GLOBAL DISTRIBUTION

Since its first report in Japan in 1884, bacterial blight has spread extensively and is now recognized as a major constraint to rice production in many rice-growing regions of the world, particularly in Asia and Africa. The global dissemination of BB reflects the historical and ongoing movement of rice germplasm, expansion of irrigated rice cultivation, and the remarkable capacity of *Xoo* to adapt to diverse ecological and agronomic environments. Today, the disease is present in nearly all continents where rice is cultivated, although its economic significance and epidemic potential vary widely among regions depending on climate, cropping intensity, varietal susceptibility, and management practices.

The widespread global distribution of BB is closely linked to the ecology of rice production systems. Warm temperatures, high relative humidity, prolonged leaf wetness, and frequent rainfall create ideal conditions for pathogen survival, multiplication, and spread. In such environments, BB often becomes endemic, persisting year after year and causing recurrent yield instability. Conversely, in regions with cooler or drier climates, the disease may occur sporadically but can still cause significant damage when favorable weather conditions coincide with susceptible crop stages.

Asia remains the epicenter of BB incidence and severity. Countries such as China, India, the Philippines, Vietnam, Thailand, Myanmar, Indonesia, Sri Lanka, and Bangladesh experience frequent and sometimes severe epidemics (Nino Liu *et al.*, 2006; Ofori *et al.*, 2025). The predominance of tropical and subtropical climates, combined with intensive rice-based cropping systems, high nitrogen input use, and the widespread cultivation of susceptible or genetically uniform varieties, creates highly favorable conditions for disease development. In many Asian countries, BB has become endemic, with predictable seasonal outbreaks that significantly constrain yield stability and necessitate continuous management interventions.

In East and Southeast Asia, irrigated lowland rice ecosystems dominate production and provide ideal conditions for BB development. Multiple cropping cycles per year, coupled with short fallow periods, facilitate pathogen carryover between seasons. In South Asia, monsoon-driven rainfall patterns often coincide with the most susceptible growth stages of the rice crop, resulting in frequent epidemics. The coexistence of traditional landraces and modern high-yielding varieties further contributes to the maintenance of diverse pathogen populations capable of overcoming deployed resistance genes.

In Africa, BB was first detected in Mali and Cameroon (Buddenhagen *et al.*, 1979). Since then, the disease has been reported from several West African countries and has progressively expanded its geographical range. The expansion of rice cultivation in Africa, particularly in irrigated and lowland systems, has provided new ecological niches for the establishment and spread of the pathogen. More recently, BB has been confirmed in East Africa, with outbreaks reported in Tanzania (Schepler Luu *et al.*, 2023) and Madagascar (Raveloson *et al.*, 2023). In June 2023, the first official report of BB caused by *Xoo* in the coastal region of Kenya was published (Nganga *et al.*, 2025). This report marked a significant milestone in the epidemiology of BB in Africa, signaling the continued expansion of the disease into new rice-producing areas.

At present, major BB outbreaks are documented in Kenya, Nigeria, Burkina Faso, Uganda, Mali, Cameroon, Tanzania, and Madagascar. The emergence and establishment of BB in Africa is of particular concern, given the increasing importance of rice as a staple food and the limited availability of resistant varieties adapted to African agroecological conditions. In many African countries, rice production is expanding rapidly to meet rising demand, often with limited disease management infrastructure. Under such circumstances, BB poses a serious threat to yield stability, farmer income, and national food security goals.

The African experience with BB highlights the role of pathogen introduction and adaptation in shaping disease distribution. Movement of seed, introduction of improved varieties without adequate resistance screening, and changing climatic conditions have likely contributed to the observed expansion of BB across the continent. Strengthened disease surveillance, regional collaboration, and targeted resistance breeding are therefore essential to limit further spread and mitigate the impact of the disease.

In Latin America, BB is comparatively less widespread and is generally reported as a sporadic disease. Localized outbreaks have been documented in countries such as Colombia and Brazil, often associated with specific environmental conditions, localized irrigated production systems, or the introduction of susceptible cultivars. Although BB has not yet reached the epidemic proportions observed in Asia or parts of Africa, its presence in Latin America underscores the potential risk of further spread under changing climatic conditions and increasing intensification of rice cultivation.

Outside the major rice-growing regions of Asia, Africa, and Latin America, BB has also been reported sporadically in other parts of the world where rice is cultivated. In temperate regions, the disease is typically limited by cooler temperatures and shorter growing seasons, but occasional outbreaks may occur under unusually favorable conditions. The global nature of rice trade and germplasm exchange means that the risk of pathogen introduction remains ever-present, reinforcing the need for vigilance even in regions where BB is currently of minor importance.

Globally, the distribution of BB reflects a complex interplay of climate, host genetics, agricultural practices, and pathogen evolution. The increasing variability of climate, characterized by extreme rainfall events and shifts in temperature regimes, is expected to further influence the spatial and temporal dynamics of BB. Changes in cropping intensity, varietal deployment, and input use may also reshape disease distribution patterns in the coming years.

A summary of the global distribution of BB is presented in Figure 2.2. This overview emphasizes that while BB is a longstanding disease, its geographical footprint continues to expand and evolve. Understanding global distribution patterns is therefore critical for guiding surveillance efforts, prioritizing resistance breeding, and

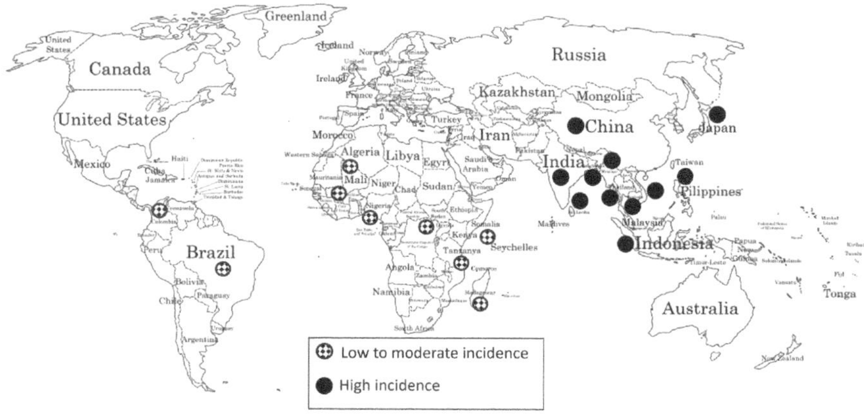

FIGURE 2.2 Global distribution of BB. Circle with dotted fill indicates low to moderate incidence while that of solid black fill indicates high incidence.

designing region-specific management strategies aimed at reducing the global burden of bacterial blight of rice.

2.3 LOSSES

Yield losses caused by BB vary widely depending on the growth stage at infection, varietal susceptibility, environmental conditions, and disease severity. Infection at early vegetative or tillering stages generally results in more pronounced yield penalties than infections occurring closer to maturity, as early damage interferes with tiller development, leaf area expansion, and photosynthetic efficiency. Under typical field conditions, yield losses are commonly reported in the range of 20% to 30%. However, under severe epidemic situations, losses may increase to 50% or more (Mew, 1987; Gnanamanickam *et al.*, 1999). Such losses can have devastating consequences for smallholder farmers, particularly in regions where rice constitutes the primary source of income and food security.

The magnitude of loss is strongly influenced by varietal response to infection. Highly susceptible varieties often show rapid lesion expansion, extensive leaf blighting, and premature drying, leading to substantial reductions in biomass accumulation and grain filling. In contrast, moderately resistant varieties may limit lesion spread but still suffer appreciable yield penalties under high disease pressure. Environmental conditions such as high relative humidity, frequent rainfall, strong winds, and optimal temperatures for pathogen multiplication further exacerbate disease severity and associated yield losses. Excessive nitrogen fertilization, commonly practiced to maximize yields, has also been shown to increase host susceptibility and intensify disease impact, thereby compounding losses.

In tropical regions of Asia, where favorable climatic conditions coincide with the cultivation of highly susceptible varieties, exceptionally high yield reductions have been documented. Reddy *et al.* (1979) reported losses of up to 74% under severe disease pressure. Similarly, in extreme situations, yield losses have been estimated to reach 65% to 71% (Rajarajeswari and Muralidharan, 2006). These figures reflect the destructive potential of BB when effective management measures are not in place. In many Asian rice-growing regions, recurring epidemics have resulted in chronic yield instability, undermining farmer confidence in susceptible high-yielding cultivars.

Recent global observations indicate that the impact of BB on yield is becoming increasingly variable and, in some regions, more severe. This trend is attributed to changing climatic patterns, including increased frequency of extreme rainfall events and prolonged periods of high humidity, which favor pathogen survival and dissemination. The intensification of rice-based cropping systems, with reduced fallow periods and increased input use, has further contributed to elevated disease pressure. In Asia and Africa, where rice cultivation is expanding into new environments, BB has emerged as a recurrent constraint capable of causing both acute epidemic losses and chronic yield suppression.

Beyond direct yield reduction, BB also affects grain quality and market value. Infected plants often produce poorly filled or chaffy grains due to impaired translocation of assimilates during grain filling. This results in lower thousand grain weight, reduced head rice recovery, and inferior milling quality. Grain discoloration and reduced uniformity may further diminish market acceptability. For farmers, these quality-related losses translate into reduced farm gate prices and diminished profitability, even in seasons when total yield losses appear moderate.

Indirect losses associated with BB are often substantial but less easily quantified. Increased expenditure on disease management, including the cost of chemical sprays, labor, and irrigation adjustments, adds to production costs without guaranteeing effective control. In severe outbreak years, farmers may be forced to replant affected fields or shift to alternative crops, resulting in additional economic losses and disruption of cropping schedules. At a community level, widespread disease outbreaks can strain local input supply systems and extension services, particularly in resource-limited settings.

At regional and national scales, the cumulative impact of BB-related losses can be significant. In countries where rice is a major contributor to national food security and rural employment, large-scale epidemics have the potential to reduce market availability, increase price volatility, and heighten food insecurity among vulnerable populations. The effects are particularly pronounced in rainfed and low-input systems, where farmers have limited capacity to absorb yield shocks or invest in alternative management options.

In Africa, where rice demand is rising rapidly and domestic production is being actively promoted, BB-related losses pose a growing threat to production targets and self-sufficiency goals. Although comprehensive loss estimates remain limited, recent outbreaks in both West and East Africa suggest that yield reductions comparable to those reported in Asia can occur under conducive conditions. The lack of widespread deployment of resistant varieties and limited access to effective management options further amplify the potential impact of the disease in these regions.

The magnitude and diversity of losses caused by BB highlight the urgent need for sustainable and durable management strategies. Chemical control options remain limited, inconsistent, and often economically unviable for smallholder farmers. Consequently, host plant resistance continues to represent the most economical, environmentally sound, and scalable approach to mitigating yield and quality losses due to BB. Understanding patterns of yield loss across environments and production systems provides a critical foundation for prioritizing research investments, guiding resistance breeding programs, and informing policy decisions.

Losses caused by BB extend beyond simple yield reduction to encompass quality deterioration, increased production costs, and broader socio-economic impacts. As rice production systems face mounting pressures from climate change, intensification, and evolving pathogen populations, the threat posed by BB is likely to persist or intensify. Comprehensive assessment of loss patterns, combined with strategic deployment of resistant varieties and integrated disease management approaches, is therefore essential to safeguard rice productivity and ensure global food security.

3 Symptoms

Bacterial blight (BB) appears in two clearly recognizable phases, the *Kresek* or seedling wilt phase and the leaf blight phase; these phases are influenced by plant age, physiological status, prevailing microclimate, inoculum load, and agronomic practices followed by farmers. Because of its ability to spread rapidly across entire fields and cause severe reductions in photosynthetic leaf area, BB remains a major limitation to sustainable rice cultivation in Asia, Africa, and parts of Latin America. A proper understanding of symptomatology is therefore essential for correct field diagnosis, quick distinction from other disorders, and effective disease management.

The symptom expression of BB closely reflects the biology of *Xoo*. The pathogen being a vascular invader enters through natural openings, hydathodes, and wounds created by mechanical injury or agronomic operations. Once it reaches the xylem, the bacterium multiplies actively, secretes virulence-associated factors (proteins/enzymes), blocks water movement, and causes gradual tissue degradation. The visible symptoms in the field are direct consequences of these internal processes. Although the disease progresses continuously from seedling to maturity, the two phases are described separately for clarity and ease of diagnosis.

3.1 *KRESEK* OR WILT PHASE

The seedling wilt or *Kresek* phase is the earliest and most severe manifestation of *Xoo* infection. It generally appears within the first week after transplanting when seedlings are under physiological stress due to uprooting and establishment in the main field. At this stage, the seedlings are highly vulnerable, especially when physical injuries are present on roots, leaf tips, or leaf sheaths as a result of rough handling during pulling, transport, or transplanting. Careful handling of seedlings before transplanting can therefore significantly reduce the occurrence of *Kresek*, since mechanical injuries provide easy entry points for the pathogen. In nurseries where bacterial inoculum persists from volunteer plants or previous crops, seedlings may also carry latent infections that become systemically expressed only after they are transplanted into the main field.

The most characteristic feature of the *Kresek* phase is the sudden onset of yellowing, loss of turgor, and rapid wilting (Figure 3.1). Seedlings that look healthy at transplanting begin to turn pale green within a short time, often within two or three days of exposure to the pathogen. The youngest leaves may first show slight drooping or curling, which may be overlooked. As infection progresses, water movement through the xylem becomes increasingly restricted because of active bacterial multiplication and production of extracellular polysaccharides that clog the vessels.

Physiologically, *Kresek* results from an almost complete disruption of the plant's water balance. Although the soil may be adequately moist or flooded, the blockage within the vascular system prevents the movement of water to the leaves. As a result, seedlings wilt and collapse as though they were experiencing drought stress. Unlike

DOI: 10.1201/9781003714590-3

FIGURE 3.1 *Kresek* or wilt phase of BB disease.

drought-related wilting, however, *Kresek* cannot be reversed. Even after repeated irrigation, affected seedlings do not recover, which reflects the fatal nature of the systemic infection.

Within three to four days, the yellowing spreads across the seedling. Leaves become limp, droop downward, and ultimately dry while still attached. The affected seedlings die completely and leave gaps in the field that cannot be compensated by later tillering.

One of the major agronomic practices that enhances *Kresek* occurrence is the clipping of seedling tops before transplanting, commonly followed for tall traditional cultivars to ease handling and ensure uniform height. This practice creates open cut surfaces at the leaf tips, which are highly susceptible to bacterial invasion. When such seedlings are transplanted into puddled fields, water facilitates the entry of the pathogen into the open tips. This situation, combined with warm temperature, high humidity, and continuous leaf wetness, creates very favorable conditions for *Xoo* establishment.

Differentiating *Kresek* from other early-season disorders is crucial for accurate diagnosis and timely management. Stem borer infestation, for example, results in the formation of dead hearts, but the affected tillers can be pulled out easily and typically reveal a hollow stem characteristic of larval feeding. In contrast, *Kresek*-affected seedlings do not show hollow stems; instead, the wilting occurs due to systemic vascular blockage caused by *Xoo* multiplication inside the xylem. Seedling wilt caused by *Rhizoctonia solani* or *Fusarium* species usually appears in localized patches and is accompanied by basal rot, crown softening, or distinct discoloration at the base of the plant. *Kresek*, however, leads to a uniform and sudden collapse of seedlings without any basal rotting or discoloration. Additionally, *Kresek* develops much earlier than the dead heart stage, which typically appears from active tillering onwards, whereas *Kresek* is specifically associated with the early seedling stage soon after transplanting.

The presence of *Kresek* in a field often indicates the potential for severe disease development later in the season. Fields that show early *Kresek* usually have uneven plant stands, delayed tillering, and reduced vigor, which make the crop more vulnerable to further blight development. This highlights the importance of early field monitoring and timely intervention, especially in regions where *Kresek* has been frequently observed.

The epidemiological significance of the *Kresek* phase lies in the very high bacterial load released from wilted seedlings. Rain splash, irrigation water flow, and mechanical operations such as weeding or gap filling can easily move the bacteria from infected seedlings to healthy ones. Even a small number of *Kresek*-affected seedlings can therefore trigger a widespread field epidemic if left unmanaged.

3.2 LEAF BLIGHT PHASE

The leaf blight phase is the most prevalent and economically important stage of bacterial blight (Figure 3.2). Unlike the sudden collapse observed in *Kresek*, leaf blight

FIGURE 3.2 Leaf blight phase of BB disease.

develops gradually and continues throughout the vegetative and reproductive growth stages. Symptoms become noticeable from early tillering and may keep expanding until the crop reaches maturity.

3.2.1 Initial Symptom Development

The earliest visible sign of leaf blight consists of tiny, translucent, water-soaked spots on the leaf blade. These often appear at the leaf tip or along the margins because these regions contain hydathodes that allow the pathogen to enter and because the leaf tip has high transpirational pull that favors bacterial movement within the xylem. Initially narrow and faint, these lesions are sometimes difficult to detect unless the leaf is viewed against strong light. These water-soaked areas indicate the beginning of xylem colonization, during which the bacteria move through the vascular system and cause leakage of cell sap into the surrounding tissues.

As the infection progresses, the lesions enlarge and may extend downward along the leaf blade as well as inward from the leaf margin toward the midrib, often simultaneously, in a wavy pattern. This characteristic directional spread, marked by irregular wavy margins, is a key feature of bacterial blight and helps distinguish it from other disorders. A distinct wavy margin appears at the junction of healthy and diseased tissue. This irregular boundary reflects the uneven progression of the pathogen through the vascular tissue and variable resistance among surrounding cells. Such a wavy pattern is regarded as a classic diagnostic feature and separates bacterial blight from many abiotic problems, which tend to show straight or uniform boundaries.

3.2.2 Progression and Expansion of Lesions

With time, the lesions expand progressively downward and may reach well beyond the middle portion of the leaf. The tissue undergoes a gradual shift in color. What begins as light green and translucent becomes yellow due to chlorophyll degradation and eventually turns straw colored when the tissue becomes completely necrotic.

In susceptible varieties, the blighted area may cover most of the leaf blade. Individual leaves may dry prematurely and hang down, causing a sharp decline in the plant's photosynthetic capacity. Tillers with multiple blighted leaves often show reduced vigor, poor tiller formation, and inefficient nutrient supply to developing panicles.

3.2.3 Identifying Tips for Typical BB Lesions

Very often, researchers lack the practical experience required to identify bacterial blight lesions solely based on field symptoms. However, there are clear and reliable visual cues that can help distinguish BB lesions with confidence. A fully developed bacterial blight lesion typically displays three distinct developmental zones, each representing a different stage of infection:

1. **Upper necrotic zone:** This region appears completely dried, brittle, and straw colored. It corresponds to the oldest part of the lesion, where the tissue has already undergone complete collapse.
2. **Middle chlorotic or watersoaked zone:** This section shows pale yellow or translucent tissue and represents the actively expanding infection front where the bacteria continue to multiply.
3. **Lower healthy green zone:** This part remains green and functional, although the advancing infection margin forms a distinct wavy boundary.

This gradient of necrosis, chlorosis, and healthy tissue is a hallmark of bacterial blight caused by *Xoo*. These three zones develop due to the directional movement of the pathogen within the xylem and the sequential collapse of tissues as the infection advances. Recognizing this gradient is especially helpful for distinguishing bacterial blight from nutrient-related disorders or from physical and chemical injuries, which do not display such a progressive and structured pattern.

3.2.4 Differentiating BB from Nutrient Deficiencies and Other Stresses

Early blight symptoms are often confused with zinc or boron deficiency because both conditions cause yellowing and drying of leaves. However, nutrient deficiencies produce more uniform discoloration over large leaf areas, without the directional, water-soaked lesions typical of bacterial blight. Nutrient deficiency symptoms usually appear simultaneously on many leaves across the plant or field, whereas bacterial blight moves progressively from the leaf tip downward and spreads from plant to plant.

Root nematode infestation can also cause streaking and leaf discoloration. However, nematode symptoms are usually irregular and not directionally progressive. Infected plants often exhibit galled or damaged roots or show specific patterns on leaf sheaths, which can help distinguish nematode injury from bacterial blight.

In advanced stages of bacterial blight, large areas of the field may appear uniformly blighted. This uniformity differentiates it from hopper burn. Hopper burn, caused by plant hoppers, occurs in distinct circular patches because the insects cluster and feed in localized groups. In contrast, bacterial blight spreads uniformly through water movement and plant-to-plant contact.

Another point of distinction is the absence of insect clusters on plants affected by bacterial blight. Hopper burn usually coincides with visible insect populations, particularly at the plant base, along with honeydew deposits. Bacterial blight symptoms, caused by a pathogen, do not show such insect-associated traces.

3.2.5 Advanced Stages and Field-Level Appearance

When the disease reaches advanced stages, large sections of the leaf dry up and droop, giving the plant a scorched appearance. The cumulative loss of functional leaf area reduces photosynthesis, limits carbohydrate production, and weakens tiller formation. This decline affects nutrient allocation and impacts grain filling, ultimately resulting in considerable yield loss.

FIGURE 3.3 Rice field with BB infection giving a blighted look.

Environmental conditions strongly influence the progression of leaf blight. Warm temperatures, high humidity, and prolonged periods of leaf wetness favor bacterial multiplication and spread. Lowland rice fields with standing water facilitate movement of the inoculum through irrigation channels. Frequent rainfall and wind help distribute the pathogen across the field. Bright sunlight increases transpiration, which pulls the bacteria deeper into the xylem and hastens lesion expansion.

High nighttime humidity promotes guttation droplets at the leaf tips, providing natural openings for pathogen entry. Dew formation on leaf surfaces supports survival and dispersal of *Xoo*. Under dry or cool conditions, symptom development may slow down, but the pathogen persists. Even moderate environmental conduciveness can cause severe disease in susceptible varieties if the inoculum load is high.

3.2.6 Impact on Crop Physiology and Yield

The impact of leaf blight extends beyond the visible leaf lesions. The reduction in functional leaf area severely restricts photosynthesis and lowers carbohydrate availability for panicle development. In severely affected plants, panicle initiation may be delayed, panicle exertion may be incomplete, and grain filling may become poor. Many grains remain partially filled or chalky because of inadequate assimilate accumulation.

In addition, infected plants experience hormonal disturbances, particularly in ethylene and abscisic acid levels, which hasten senescence. In extreme cases, premature drying of leaves occurs even before panicle emergence, leading to significant yield reduction. The overall physiological stress caused by blight increases the vulnerability of plants to other biotic and abiotic stresses, further compounding the damage.

Given these consequences, a clear and detailed understanding of symptomatology is essential for early detection, proper field monitoring, and informed decision-making in disease management.

4 The Pathogen

4.1 TAXONOMIC POSITION

Xanthomonas oryzae pv. *oryzae* is placed within a well-resolved taxonomic framework considering both classical phenotypic systematics and current molecular phylogenetic revisions. The genus *Xanthomonas* was first described by Dowson in 1939 (Dowson, 1939, 1943) and has undergone extensive taxonomic refinement due to advances in DNA–DNA hybridization studies, multilocus sequence analysis, average nucleotide identity, and whole genome phylogenetics. Earlier classifications placed *Xanthomonas* within Proteobacteria, but the phylum has been renamed Pseudomonadota following updated bacterial taxonomy supported by genomic phylogenomics.

4.1.1 Taxonomic Hierarchy

- **Domain:** Bacteria
- **Phylum:** Pseudomonadota (previously Proteobacteria)
- **Class:** Gammaproteobacteria
- **Order:** Xanthomonadales
- **Family:** Xanthomonadaceae
- **Genus:** *Xanthomonas* (Dowson, 1939)
- **Species:** *X. oryzae* (Ishiyama, 1922; Swings *et al.*, 1990)
- **Pathovar:** *X. oryzae* pv. *oryzae* (Ishiyama, 1922; Swings *et al.*, 1990)

4.1.2 Taxonomic Significance and Characteristic Features

Members of the genus *Xanthomonas* share several diagnostic traits:

1. **Production of characteristic yellow xanthomonadin pigment**: This membrane-bound, brominated aryl polyene pigment protects cells from photooxidative damage and is unique to *Xanthomonas*, making it an important chemotaxonomic marker (Poplawsky *et al.*, 2000).
2. **Presence of a distinctive type III secretion system (T3SS):** Although many plant pathogenic bacteria possess T3SS, the repertoire of transcription activator-like effectors (TALEs) is a hallmark of *X. oryzae* pathovars (Bogdanove *et al.*, 2010).
3. **Unique lipid A and LPS composition**: Fatty acids such as 3-hydroxy fatty acids and methyl-branched fatty acids are characteristic of *Xanthomonadaceae* family members.
4. **DNA G+C content**: Typically ranges from 64% to 69%, which is higher than that of many other Gammaproteobacteria and serves as a molecular signature for the genus.

DOI: 10.1201/9781003714590-4

4.1.3 Differentiation among *Xanthomonas* Pathovars

Within the genus, several agriculturally important pathovars exist, including:

- *Xanthomonas citri* subsp. *citri* (Citrus canker),
- *Xanthomonas campestris* pv. *campestris* (Black rot of crucifers),
- *X. citri* subsp. *glycines* (Bacterial pustule of soybean),
- *X. citri* subsp. *fuscans* (Fuscous bean blight).

These pathovars differ in:

- xanthomonadin intensity,
- extracellular polysaccharide production,
- metabolic versatility,
- host specificity,
- colony morphology,
- pathogenicity genes, including TALE complements.

Xoo is distinct among these due to the presence of *rich TALE repertoires*, and *slow growth on agar media* compared to more rapidly growing pathovars like *X. citri.*

4.2 PHENOTYPIC CHARACTERS

4.2.1 Colony Characteristics

- **Pigment:** Mature colonies of 72–96 hours old on YGCA medium (10 g yeast extract, 10 g D-glucose, 20 g calcium carbonate, and 20 g agar) appear as light-yellow or pale yellow colored that shines and glistens under direct light. The pigment is not diffusible; therefore, the medium surrounding the colonies remains unchanged in color. This property differentiates *Xoo* from certain *Pantoea* species, which produce bright diffusible yellow pigments.

  The yellow pigment produced by *Xoo* can be extracted in methanol. Spectrophotometric scanning at a wavelength of 440 nm confirms the identity of xanthomonadin. The pigment's resistance to 1% sodium hydroxide further distinguishes *Xanthomonas* species from *Pseudomonas* species. Quantification of pigment production also provides insights into virulence and UV-resistance traits, which may vary among isolates.
- **Texture:** Mucoid—due to extracellular polysaccharide xanthan. The degree of mucoidity varies among strains depending on xanthan production genes such as *gumB*, *gumD*, and *gumK*.
- **Topography:** Convex-dome shaped.
- **Shape and margin:** Circular with smooth margins.
- **Size:** 2.5–5 millimeters in diameter.

4.2.2 Key Differentiation from Other Yellow Pigmented Bacteria

4.2.2.1 Differentiation from *Pantoea*

- *Pantoea* colonies are bright yellow or orangish.

- Growth is rapid, visible within 24–36 hours.
- Colonies are smooth but not mucoid.
- Produces indole and is often positive for many carbohydrate utilization tests.

4.2.2.2 Differentiation from *Sphingomonas*

- Colonies are dry and dull.
- No mucoid texture.
- Growth is extremely slow but still earlier than *Xoo* in many cases.
- Produces glycosphingolipids, giving a characteristic dry texture.

4.2.2.3 Differentiation from Other *Xanthomonas* Pathovars

- *X. citri*: Deep yellow colonies and faster growth
- *X. axonopodis*: Produces abundant xanthan and darker yellow pigmentation
- *X. campestris*: Strongly mucoid with thick xanthan slime
- *X. citri* subsp. *fuscans*: Produces brown to fuscous pigments

Xoo stands out due to its light yellow, glistening, non-diffusible pigment and slow but steady growth.

4.2.3 Growth Characteristics on Different Media

4.2.3.1 YGCA

On YGCA (yeast extract glucose calcium carbonate agar), *Xoo* forms light yellow colonies with glistening surfaces. The colonies are circular and convex with smooth margins. The calcium carbonate in the medium produces a subtle, opaque white background that enhances the visibility of the colonies. Pigment does not diffuse into the agar, and therefore the surrounding medium remains unchanged.

Colonies reach a diameter of 2.5–5 millimeters within 72–96 hours. Colonies that appear earlier than 48 hours are usually contaminants such as *Pantoea*, *Sphingomonas*, *Bacillus*, or fast-growing saprophytic yellow bacteria. Marking early colonies with a glass marker helps in avoiding confusion during selection.

4.2.3.2 Growth on PSA

On PSA (peptone sucrose agar), *Xoo* produces more mucoid colonies due to high sucrose content that stimulates EPS synthesis. Colonies are pale yellow and sticky with a distinct convex elevation (Figure 4.1). The medium allows clearer differentiation because *Pantoea* forms bright yellow to orange colonies, and *Sphingomonas* forms dry yellow colonies without mucoidy.

4.2.3.3 Growth on GYCA and SX Agar

GYCA (glucose yeast calcium agar) supports moderate growth with slightly reduced mucoidy compared to PSA. SX (Sucrose–Xanthomonas agar, a semi-selective medium commonly used for the isolation of Xanthomonas species) agar, developed specifically for *Xanthomonas* species, enhances pigment expression. *Xoo* colonies on SX agar appear distinctly yellow but remain non-diffusible. In contrast, *X. citri* and *X. campestris* often show deeper yellow or golden pigmentation, which helps in differentiation.

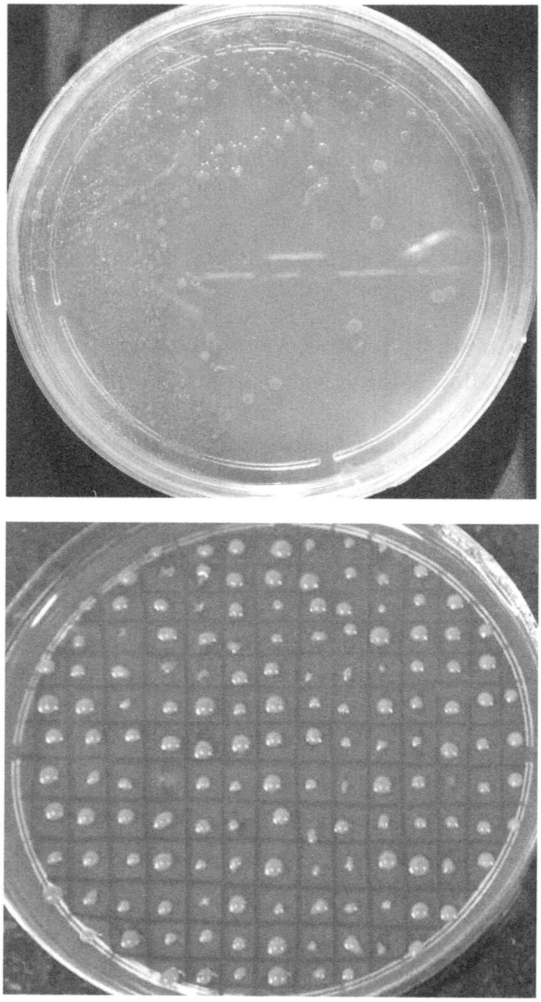

FIGURE 4.1 Colonies of *Xoo* on NSA media. Growing colonies (above) and fully grown colonies of *Xoo* strains (below).

4.3 KEY PHYSIOLOGICAL AND BIOCHEMICAL PROPERTIES

4.3.1 Gram Reaction and Cell Wall Chemistry

The bacterium is Gram negative, retaining the counterstain in Gram staining. Its cell wall consists of a double-unit membrane, with permease proteins embedded fully or partially within the phospholipid bilayer, consistent with the fluid-mosaic model. Distinctive features of the cell wall include a single-layered peptidoglycan, lipopolysaccharides, lipoproteins, and a periplasmic space (Figure 4.2).

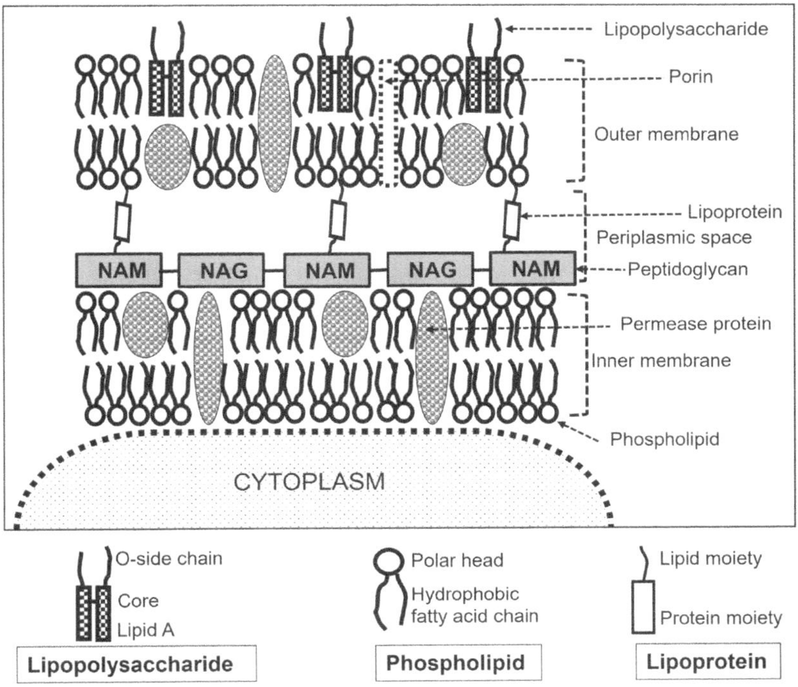

FIGURE 4.2 *Xoo* cell wall with characteristic feature.

4.3.2 KOH Test

The KOH test is a simple and rapid method used to differentiate between Gram negative and Gram-positive bacteria. A positive reaction is indicated by the development of a sticky or thread-like consistency when bacterial colonies are mixed with a 3% KOH solution. This sticky reaction occurs because the alkaline KOH disrupts the weaker cell wall structure of Gram negative bacteria, causing the outer membrane to lyse. As the cells break down, the bacterial DNA is liberated into the solution, imparting a characteristic viscous, stringy appearance when lifted with an inoculation loop or toothpick.

In contrast, Gram-positive bacteria possess a thick peptidoglycan cell wall that resists lysis in the presence of KOH, so no sticky reaction is observed. Thus, the KOH test serves as a confirmatory assay for Gram reaction, complementing traditional Gram staining.

To perform the test, a small portion of a fresh colony is thoroughly mixed with a drop of 3% KOH on a clean glass slide. After mixing for 30–60 seconds, the inoculating loop is lifted:

- If a sticky thread or string is formed, the organism is Gram negative.
- If the mixture remains watery and no stringiness develops, the organism is Grampositive.

This simple test provides a quick, reliable, and cost-effective means to distinguish Gram reaction, especially in routine bacteriological diagnostics.

4.3.3 Differentiating Physiological and Biochemical Features

Xoo tolerates moderate osmotic stress but is highly sensitive to UV and desiccation. This explains its limited survival on exposed surfaces and dependence on plant tissues for persistence. The pigment does not provide UV protection, unlike carotenoid pigments found in *Sphingomonas.*

Xoo is aerobic and oxidizes glucose through the Entner–Doudoroff pathway. The organism lacks a functional oxidative fermentative glucose system, which makes the oxidation–fermentation test negative. Enzymes that degrade plant cell walls, such as cellulases and pectinases, are either absent or produced in negligible quantities, which explains the lack of tissue maceration in rice leaves.

Xoo grows optimally at 28 °C and neutral pH. High sucrose stimulates xanthan production while low nitrogen conditions reduce growth. The bacterium cannot utilize many simple amino acids like L-alanine as a sole carbon source, which helps in diagnostic separation from *Pantoea.*

The following biochemical features are useful in differentiating *Xoo* from morphologically similar yellow-pigmented bacteria.

- **Oxidase:** Negative
- **Catalase:** Positive
- **L-alanine utilization:** Negative
- **Two ketogluconate production:** Negative
- **Fluorescence on King B medium:** Negative
- **Nitrate reduction:** Negative
- **Acetoin production (Voges Proskauer):** Variable
- **Oxidation fermentation test:** Negative
- **Gelatin hydrolysis:** Negative
- **Starch hydrolysis:** Negative
- **Growth on 0.2% vitamin free casamino acids:** Negative
- **Peptonization of litmus milk:** Negative

4.3.4 Biolog Profiles

Automated Biolog MicroPlates™ facilitate high-throughput assessment of carbon-source utilization patterns, generating distinct metabolic fingerprints. These metabolic profiles effectively distinguish *Xoo* from other rice-associated bacteria such as *Pantoea* and *Sphingomonas* species, contributing to precise identification and characterization of the pathogen.

Biolog GN2 and GENIII assays show the following typical *Xoo* characteristics:

- Limited carbohydrate utilization
- Negative for many amino acid and organic acid substrates

- Sensitivity to increased osmolarity and inhibitory compounds
- Clear separation from *Pantoea*, which shows broad metabolic activity

The metabolic conservatism of *Xoo* is consistent with its specialized lifestyle as a vascular pathogen.

4.3.5 FAME Profile

Fatty acid methyl ester analysis is a reliable method for differentiating *Xoo* from related bacteria and thus constitutes a dependable chemotaxonomic marker for species identification.

FAME profile is having:

- High proportion of 3-hydroxy fatty acids,
- Presence of iso $C_{15:0}$ and anteiso $C_{15:0}$,
- Dominance of $C_{16:0}$ and $C_{18:1}$ omega7c.

Pantoea lacks significant hydroxy fatty acids, while *Sphingomonas* contains unique long-chain sphingolipids.

4.4 CELL STRUCTURE AND GENERAL MORPHOLOGY

Cells of *Xoo* are straight rods that occur singly and rarely in short pairs. The cells have dimensions typically ranging from 0.50 to 0.75 μm in width and 1.30 to 2.15 μm in length. They appear as Gram negative rods with smooth surfaces and clearly visible cell boundaries under bright field microscopy. The bacterial cells are non-spore–forming and exhibit an even cytoplasmic distribution without any inclusion bodies that can be observed under standard light microscopy.

Under phase contrast microscopy, *Xoo* cells appear refractile with a central light region due to their aqueous cytosol. In wet mount preparations using early log phase cultures, active swimming is observed because the cells possess a single polar flagellum that drives a darting type of motility. In older cultures, motility is reduced, and cells often appear aggregated due to the abundant production of extracellular polysaccharides.

4.4.1 Ultrastructure Based on EM Studies

Transmission electron microscopy of *Xoo* shows:

- A clearly defined outer membrane with a thin peptidoglycan layer,
- Evenly distributed ribosomes, giving a granular cytoplasmic appearance,
- No intracellular vacuoles,
- Presence of periplasmic space that is more prominent in older cultures,
- A polar flagellum inserted into a deep invagination of the outer membrane.

Scanning electron microscopy reveals smooth-surfaced rod-shaped cells that often occur embedded in an EPS matrix. Under nutrient-rich conditions, the cells grow as microcolonies forming dense aggregates, whereas on minimal media they remain more dispersed.

4.4.2 Motility and Flagellar Architecture

Xoo is monotrichous, possessing a single polar flagellum. Transmission electron microscopy reveals a long and flexible flagellar filament generally ranges from about 5 to 10 μm in length (approximately 2 to 8 times the length of the bacterial cell.). The flagellum is composed of flagellin subunits encoded by the *fliC* gene, which has been extensively used in phylogenetic studies among *Xanthomonas* species (Zhang *et al.*, 2015). Besides, the antigenic property of flagella is primarily attributed to conserved epitopes within the flagellin protein, particularly the N-terminal flg22 region that is recognized by host receptors. The fliC-encoded flagellin acts as a potent elicitor of plant innate immunity, triggering pattern-triggered immune (PTI) responses upon perception by receptors such as FLS2. Variations in flagellin amino acid sequences contribute to antigenic diversity and may help the pathogen evade host recognition. Furthermore, expression of fliC is tightly regulated within a hierarchical flagellar gene network involving regulators such as FleQ, RpoN2, and FliA. Thus, both structural and regulatory features of flagellin play crucial roles in motility, antigenicity, and host–pathogen interactions in Xanthomonas.

The motility of *Xoo* not only is important for dispersal but also contributes to early events of host colonization. The flagellar motor functions efficiently at neutral pH and moderate salt conditions, while extreme osmotic and oxidative stresses reduce motility significantly. Unlike many soft rot pathogens, *Xoo* does not exhibit swarming motility on semi-solid media, which is an important diagnostic feature when differentiating it from certain saprophytic yellow bacteria.

4.4.3 Surface Polymers and Structural Components

Xoo synthesizes abundant extracellular polysaccharides (EPS), primarily xanthan, encoded by the *gum* gene cluster. Xanthan contributes to colony mucoidy, leaf streak appearance, and clogging of xylem vessels in the host plant. In culture, EPS production increases during the stationary phase, leading to slimy and sticky colonies that retain moisture longer than colonies of *Pantoea* or *Sphingomonas.*

Lipopolysaccharides (LPS) form a major component of the outer membrane and consist of lipid A, core oligosaccharide, and O antigen chains. *Xoo* LPS contributes to virulence by suppressing plant defense responses and is structurally distinct from LPS of *Sphingomonas*, which contains glycosphingolipids instead of typical lipid A moieties. This distinction is diagnostically significant because *Sphingomonas* colonies appear dry and non-mucoid due to the absence of substantial EPS production.

5 Isolation and Pure Culture Establishment of *Xoo*

Isolation of *Xoo* is a fundamental requirement for pathological, molecular, physiological, and epidemiological investigations. Even though molecular diagnostics and immunoassays have greatly improved the speed of pathogen detection, isolation and culture of the bacterium remain indispensable for confirming pathogenicity, studying host–pathogen interactions, assessing virulence, and maintaining reference stocks. The process relies on exploiting the ability of viable *Xoo* cells present in symptomatic rice tissues to grow under controlled laboratory conditions and produce characteristic yellow, mucoid colonies resulting from xanthomonadin pigment and extracellular polysaccharide (EPS) secretion.

5.1 GROWTH MEDIA FOR SELECTIVE ENRICHMENT AND COLONY DEVELOPMENT

Isolation typically begins by surface-sterilizing infected leaf bits in 70% ethanol or dilute sodium hypochlorite, followed by streaking on nutrient media supporting *Xoo* growth. Wakimoto medium, composed of peptone, sucrose, calcium nitrate, disodium phosphate, and agar at neutral pH, is frequently used because it encourages the development of slow-growing, yellow, mucoid colonies typical of the pathogen. Other widely used media include Yeast Glucose Calcium Carbonate Agar (YGCA), where *Xoo* produces bright yellow, mucoid colonies within 48–72 hours, and Peptone Sucrose Agar (PSA) or Yeast Dextrose Calcium Carbonate Agar (YDCA), which similarly support characteristic colony formation. These media not only enable the growth of *Xoo* but also help differentiate it from faster-growing saprophytes based on the non-diffusible yellow pigmentation and the distinctive mucoid texture.

5.2 ISOLATION AND CULTURE ESTABLISHMENT

Successful isolation depends heavily on selecting appropriate plant material. The advancing margin of a fresh bacterial blight lesion, where the transition from healthy to diseased tissue is clearly visible, contains the highest population of actively multiplying *Xoo* cells. Small sections of approximately 2–3 mm width, including both the lesion and adjacent healthy tissue, are excised using a sterile scalpel or scissors. These pieces are surface sterilized—usually by immersing them briefly in 0.15% sodium hypochlorite or 70% ethanol—and rinsed thoroughly in sterile distilled

DOI: 10.1201/9781003714590-5

water to remove residual disinfectants. Excess moisture is removed by blotting on sterile tissue paper.

The sterilized leaf fragments are then placed in a small volume of sterile distilled water and allowed to stand for about 10–15 minutes. During this period, bacterial cells and extracellular ooze diffuse from the vascular bundles into the surrounding water, creating a slightly turbid suspension suitable for plating. Using a sterile inoculating loop, a portion of this suspension is streaked across the surface of PSA, YGCA, or Wakimoto agar plates in a manner that facilitates separation of individual colonies. Plates are incubated in an inverted position at 28 ± 2 °C to prevent condensation from falling onto the agar surface. Because *Xoo* grows relatively slowly, colonies typically begin to appear only after 72–96 hours. In contrast, many saprophytic bacteria, yeasts, and other yellow-pigmented organisms may emerge within the first 24–48 hours. These early-appearing colonies are generally not *Xoo* and should be visually marked to avoid confusion during subsequent observations. Careful inspection after three to four days generally reveals the authentic *Xoo* colonies, which are round, smooth, convex, shiny, and bright yellow in appearance.

Once suspected colonies are identified, they must be purified by repeated restreaking onto fresh plates until a uniform colony morphology is obtained. Isolates confirmed through repeated subculturing are then subjected to phenotypic and molecular confirmation to ensure authenticity and purity.

5.3 MORPHOLOGICAL AND CULTURAL CONFIRMATION

Pure colonies of *Xoo* on PSA or YGCA are circular, convex, smooth-surfaced, and light to bright yellow, a color arising from the xanthomonadin pigment that remains non-diffusible in the medium. After extended incubation, colonies become mucoid due to abundant EPS secretion, further distinguishing them from many saprophytes. Microscopic observation of Gram-stained smears from young colonies shows short, rod-shaped, Gram negative cells. A positive KOH solubility reaction confirms the Gram negative nature of the isolates. Under phase-contrast microscopy, *Xoo* cells reveal a rapid, darting motility typical of organisms possessing a single polar flagellum.

Additional cultural characteristics include the absence of fluorescence when grown on King's B medium and the inability to hydrolyze starch, liquefy gelatin, or reduce nitrate—features that support differentiation from related bacteria. Unique fatty acid methyl ester (FAME) profiles and Biolog metabolic fingerprints further contribute to the confirmation of identity.

5.4 PATHOGENICITY TESTING FOR FINAL CONFIRMATION

Even after phenotypic and molecular identification, pathogenicity tests are essential to satisfy Koch's postulates. Susceptible rice varieties such as TN1 or IR24 are grown to the tillering stage under controlled conditions. A bacterial suspension is prepared by scraping growth from a 48–72-hour plate culture and suspending the cells in sterile water to obtain approximately 10^8 CFU mL^{-1} (equivalent to $OD_{600} \approx 0.3$). Pathogenicity is assessed using the widely adopted leaf-clipping method, wherein the tips of fully expanded leaves are clipped using scissors dipped in the bacterial suspension. Plants are then maintained under high humidity and moderate temperature to promote infection. Typical blight symptoms—initially water-soaked streaks along the leaf margin that later turn yellow and expand downward—appear within 10–14 days. Re-isolation of the bacterium from these lesions confirms the pathogenic nature of the isolate as *Xoo*.

5.5 LONG-TERM STORAGE AND CULTURE MAINTENANCE

Once fully confirmed, pure cultures should be preserved using methods that ensure long-term viability and stability of pathogenic traits. Cryopreservation in 50% glycerol stocks stored at –80 °C is highly reliable for extended preservation. Lyophilization offers another stable option for long-term storage. For routine laboratory activities, cultures may be maintained on YGCA or PSA slants at refrigeration temperature (about 4 °C), although frequent subculturing should be avoided as it may lead to loss of pigmentation or virulence.

5.6 PRECAUTIONS AND CRITICAL NOTES

Isolation of *Xoo* demands strict attention to aseptic technique, careful selection of sampling tissue, and patient observation of colony development. Overexposure to disinfectants during surface sterilization can kill cells at the lesion margin, significantly reducing isolation success. Conversely, insufficient sterilization may result in plates being overgrown with contaminants. During incubation, humidity should be maintained to prevent the drying of agar, as desiccation hampers *Xoo* growth. Most importantly, subculturing should be minimized to maintain the physiological and pathogenic integrity of isolates, particularly when they are intended for virulence studies, race characterization, or genomic analyses.

Workflow: Isolation and Pure Culture Establishment of *Xoo*

Step 1. Sample Selection

- Choose *fresh BB lesions* showing an *advancing margin* between healthy and diseased tissue.
- Avoid old, dry, or secondary-infected lesions.

Step 2. Surface Sterilization of Leaf Tissue

1. Cut 2–3 mm lesion pieces, including healthy margin.
2. Immerse in:
   - o *70% ethanol for 30–45 sec* or
   - o *0.15% sodium hypochlorite for 30–45 sec*
3. Rinse *2–3 times* with sterile distilled water.
4. Blot dry on sterile tissue.

Step 3. Bacterial Ooze Extraction

- Place sterilized pieces in 200–300 µL sterile distilled water.
- Allow *10–15 min* for bacterial exudation and formation of a turbid suspension.

Step 4. Primary Isolation (Plating)

1. Streak the ooze suspension on one or more media:
   - o **Wakimoto medium**
   - o **YGCA/YDCA**
   - o **PSA**
2. Incubate plates *inverted* at *28 ± 2 °C.*
3. Note colony timing:
   - o Non-*Xoo* contaminants: *appear within 24–48 h*
   - o Typical *Xoo* colonies: *bright yellow, mucoid, appear 72–96 h*

Step 5. Identification of Suspected *Xoo* Colonies

- Look for colonies that are:
  - o Round, smooth, convex
  - o Shiny, light to bright yellow
  - o Non-diffusible pigment (xanthomonadin)
  - o Mucoid due to EPS (with longer incubation)

Step 6. Purification of Colonies

- Pick suspected *Xoo* colonies and *restreak onto fresh PSA/YGCA plates.*
- Repeat until *uniform single-colony morphology* is obtained.

Step 7. Preliminary Confirmation Tests

Morphological and Microscopy

- Gram stain: *Gram negative rods*
- KOH test: *positive*
- Motility: *rapid polar-flagellar darting*

Cultural Traits

- No fluorescence on King's B medium
- Slow growth compared to saprophytes
- Characteristic yellow, mucoid colonies

Step 8. Additional Biochemical/Phenotypic Confirmation

(Details already in Chapter 4)

- Starch hydrolysis: *negative*
- Gelatin liquefaction: *negative*

- Nitrate reduction: *negative*
- Biolog metabolic fingerprinting
- FAME fatty acid profile signature

Step 9. Pathogenicity Test (Koch's Postulates)

1. Prepare inoculum (48–72 h culture) at *~10^8 CFU/mL* ($OD_{600} \approx 0.3$).
2. Clip leaf tips of susceptible rice (TN1/IR24) using scissors dipped in inoculum.
3. Maintain high humidity.
4. Observe BB symptoms in *10–14 days*.
5. Re-isolate from lesion to confirm identity.

Step 10. Long-Term Preservation

- *–80 °C glycerol stocks (50%)*—most reliable
- *Lyophilization* for long-term archival
- *Slants at 4 °C* for short-term use (avoid frequent subculturing)

Critical Precautions

- Avoid over-sterilization (kills lesion margin bacteria).
- Avoid under-sterilization (heavy contamination).
- Maintain proper humidity during incubation (to prevent agar drying).
- Minimize subcultures to preserve *pigmentation, virulence, and physiological traits*.

6 Detection and Diagnosis

Early and precise detection of *Xoo* is vital to the timely implementation of effective management strategies, ultimately contributing to sustainable rice production and global food security. Accurate and prompt diagnosis supports all levels of disease management, ranging from seed health confirmation, epidemic surveillance, and planting material certification to the enforcement of quarantine regulations.

Traditional diagnostic methods such as symptom observation, pathogen isolation, and biochemical characterization continue to play an indispensable role in *Xoo* identification. However, these techniques are often time-consuming, labor-intensive, and may yield inconclusive results. In recent years, advances in molecular biology, immunodiagnostics, nanotechnology, and artificial intelligence have revolutionized plant disease diagnostics, offering rapid, specific, and field-deployable tools. This chapter presents an overview of both conventional and modern diagnostic approaches for *Xoo*, focusing on their sensitivity, specificity, and applicability under field and laboratory conditions.

6.1 CONVENTIONAL DIAGNOSTIC METHODS

6.1.1 Symptom-Based Diagnosis

Typical BB symptom appears about 10–14 days after infection and differ between the leaf blight and *kreshek* phases. On fully expanded leaves, initial water-soaked lesions develop along the margins, later enlarging into blighted lesions with wavy edges. In severe infections, leaves dry from the tip downward, producing a scorched appearance. At the seedling stage, systemic infection results in complete wilting and collapse of the plant, a condition known as *kresek*.

Differentiating BB symptoms from those of other diseases, such as sheath blight, or from nutrient deficiencies and mechanical injuries, is essential. Careful field observation, particularly of lesion margins and the presence of bacterial ooze, greatly aids preliminary identification.

6.1.2 Colony Morphology and Pigment Analysis

Colonies of *Xoo* produce a distinctive canary-yellow pigment that can be extracted in methanol. Spectrophotometric scanning at a wavelength of 440 nm confirms the identity of xanthomonadin. The pigment's resistance to 1% sodium hydroxide further distinguishes *Xanthomonas* species from *Pseudomonas* species. Quantification of pigment production also provides insights into virulence and UV-resistance traits, which may vary among isolates.

DOI: 10.1201/9781003714590-6

6.1.3 Biochemical Profile or Biolog Phenotyping–Based Diagnosis

The detailed biochemical and Biolog phenotypic characterization of *Xanthomonas* species has already been comprehensively addressed in Chapter 4 (Sections 4.3.3 and 4.3.4). In this chapter, only key diagnostic considerations are highlighted:

- Biochemical reaction patterns—including starch hydrolysis, gelatin liquefaction, catalase activity, and oxidative metabolism—serve as supportive indicators for confirming *Xanthomonas* identity.
- Biolog GN2/GN3 microplate profiling provides a high-resolution metabolic fingerprint based on carbon-source utilization, enabling differentiation of closely related *Xanthomonas* pathovars.
- While not species-exclusive, these phenotypic signatures strengthen presumptive identification when used in combination with morphological and molecular assays.

6.1.4 FAME Analysis–Based Diagnosis

As described earlier in Chapter 4 (Section 4.3.5), Fatty Acid Methyl Ester (FAME) profiling offers a chemotaxonomic approach for distinguishing *Xanthomonas* taxa. Only the essential points are summarized here:

- FAME analysis relies on gas chromatography to quantify cellular fatty acid composition, producing a species-level "fatty acid signature."
- Characteristic branched-chain fatty acids—especially iso- and anteiso-C15:0 and C17:0—provide key diagnostic markers supportive of *Xanthomonas* identification.
- FAME profiles are particularly useful for confirming identity within culture collections or reference laboratories, complementing phenotypic and molecular diagnostics.
- Because environmental and growth-phase factors can influence profiles, FAME results are best interpreted alongside standardized protocols and authenticated reference libraries.

6.1.5 Bacterial Ooze Test

The bacterial ooze test is a simple, rapid, and highly reliable diagnostic tool used to confirm the presence of BB in rice (Figure 6.1). It is particularly valuable for preliminary diagnosis as it can be easily conducted both in the field and in the laboratory without the need for sophisticated instruments. This test enables farmers, plant pathologists, and researchers to distinguish *Xoo* infections from other biotic or abiotic disorders that may produce similar foliar symptoms.

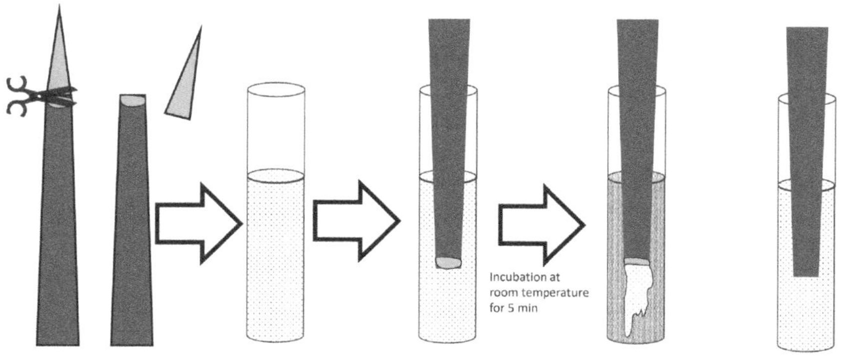

FIGURE 6.1 Bacterial ooze test.

6.1.5.1 Principle of the Test

The principle of the bacterial ooze test is based on the unique ability of *Xoo* to colonize the vascular bundles, particularly the xylem vessels, of infected rice leaves. The pathogen multiplies extensively within these vessels, producing large quantities of bacterial cells and extracellular polysaccharides (EPS). When the infected leaf is cut and its end is immersed in clean water, the bacterial mass along with EPS exudes from the cut surface, forming a visible milky or cloudy stream. This characteristic exudation serves as a distinct visual indicator of *Xoo* infection and can be easily observed with the naked eye, even under field conditions.

6.1.5.2 Materials Required

To perform the bacterial ooze test, the following materials are required:

- Rice leaves exhibiting typical BB symptoms (preferably at the active lesion stage)
- Sterile scissors or a sharp blade for cutting leaf samples
- Sterile test tubes, watch glasses, or small beakers
- Sterile distilled or clean water (preferably free of chlorine or contaminants)
- Forceps (optional, for handling samples)
- A light source or white background for clear visualization of ooze

6.1.5.3 Working Protocol

1. **Collection of sample**: Select rice leaves showing characteristic bacterial blight lesions, typically starting from the leaf tip or margin. The lesions are water-soaked in the early stages, later turning yellowish-white with wavy edges. For best results, choose leaves from actively progressing infections rather than completely dried ones, as fresh lesions contain higher bacterial loads.

2. **Preparation of sample**: Using a sterile blade or scissors, cut a small portion of the diseased leaf (about 3–5 cm long). Ensure that the cut passes through the junction of healthy and diseased tissue, as the bacterial concentration is generally highest near the advancing lesion margin. Handle the sample carefully to avoid cross-contamination between leaves.
3. **Setting up the test**: Place the freshly cut end of the leaf segment into a sterile test tube, watch glass, or small beaker containing 5–10 mL of sterile distilled water. The cut surface should remain submerged in water but not crushed or fully immersed. Multiple samples can be tested simultaneously, provided that separate containers are used to prevent cross-contamination.
4. **Observation**: Allow the setup to remain undisturbed for 5–15 minutes. To enhance visibility, hold the container against natural light or place it on a white background. In positive cases, a distinct cloudy or milky stream can be observed emanating from the cut leaf ends, gradually diffusing into the surrounding water. This visible streaming confirms the presence of bacterial exudate released from the vascular tissues.

6.1.5.4 Interpretation of Results

- **Positive result:** A clear, milky, or turbid streaming from the cut end of the leaf indicates the presence of *Xoo*. The ooze represents bacterial cells and EPS released into the water column. This observation is diagnostic of bacterial blight and distinguishes it from fungal or abiotic disorders that do not produce such streaming.
- **Negative result:** The absence of visible streaming suggests that *Xoo* is not present or that the bacterial population within the tested tissue is too low to produce detectable ooze. Negative results may also occur when leaves are too old or desiccated, or when symptoms are caused by nutrient deficiencies or insect injuries rather than bacterial infection.

6.1.5.5 Precautions and Critical Notes

1. Freshly infected leaves yield the most reliable results; old or completely necrotic tissues often give false negatives due to reduced bacterial activity.
2. Use only sterile instruments and containers to avoid contamination from other microorganisms.
3. Conduct the test under moderate temperature conditions (around 25–30 °C) for optimal bacterial activity.
4. If the water becomes cloudy without distinct streaming, repeat the test using fresh samples to rule out environmental contaminants.
5. For confirmatory diagnosis, positive samples can be subsequently cultured on selective media such as Wakimoto or PSA to isolate pure colonies of *Xoo*.

The bacterial ooze test remains one of the most practical and cost-effective diagnostic tools for detecting *Xoo*, especially in field conditions where advanced facilities are unavailable. It provides quick visual confirmation of bacterial blight within minutes, supporting timely disease management decisions. This rapid screening method is widely used by extension workers and plant health inspectors during disease surveillance, seed health certification, and epidemiological studies.

In integrated diagnostic workflows, the bacterial ooze test serves as the first-line screening method before moving to confirmatory laboratory analyses such as biochemical characterization or molecular assays. Its simplicity, reproducibility, and visual clarity make it an indispensable part of the diagnostic toolkit for rice pathologists and field practitioners.

6.2 *XOO*-SPECIFIC DNA MARKERS

The development and application of DNA-based molecular markers have revolutionized plant disease diagnostics by enabling rapid, sensitive, and specific detection of pathogens directly from infected plant material. In the case of bacterial blight of rice, genome sequencing of multiple *Xoo* strains has facilitated the identification of unique genetic regions that serve as reliable molecular signatures for this pathogen. These specific DNA markers form the basis for precise diagnosis, strain differentiation, and population surveillance, thereby complementing traditional isolation and culturing techniques.

6.2.1 Principle

The basic principle behind the use of Xoo-specific DNA markers is the amplification of pathogen-specific DNA sequences using polymerase chain reaction (PCR). Through comparative genomic studies, regions of the *Xoo* genome that are conserved within the species but absent in other *Xanthomonas* pathovars or related bacterial genera are identified. Primers designed from these regions ensure selective amplification of *Xoo* DNA only, thereby reducing the likelihood of false positives or cross-reactivity.

When a plant sample suspected of harboring *Xoo* is subjected to PCR using such specific primers, the appearance of an amplicon of known size confirms the presence of the pathogen. This molecular approach provides a rapid alternative to time-consuming cultural and biochemical methods and is particularly valuable when bacterial populations in plant tissue are low or when symptoms are unclear.

6.2.2 Primer Design and Development

Designing reliable primers requires a careful bioinformatic comparison of whole-genome sequences from multiple *Xoo* strains and closely related *Xanthomonas* species. The selection criteria for suitable target regions include the following:

1. **Specificity**—the chosen sequence must be unique to *Xoo* and absent from other pathovars.
2. **Conservation**—it should be conserved across diverse *Xoo* isolates from different geographical regions.

3. **Stability**—it should not be located in highly mutable regions such as repetitive elements, unless those repeats are unique and stable within *Xoo*.
4. **Amplification efficiency**—the amplicon size should be optimized (generally 150–250 bp) to ensure efficient amplification in both conventional and real-time PCR platforms.

Modern primer design utilizes bioinformatic tools such as BLAST, Primer3, and OligoAnalyzer, which assess sequence specificity, melting temperature, GC content, and secondary structure formation. These computational checks ensure that primers selectively target *Xoo* without amplifying non-target DNA from host plants or other microorganisms.

6.2.3 Examples of *Xoo*-Specific DNA Markers

Several well-characterized DNA markers have been successfully employed for the detection of *Xoo* in both laboratory and field settings. Some of the widely used examples include the following:

1. **IS1113 Insertion Sequence–Based Marker**
   - **Primers:** TXT (5′ GTCAAGCCAACTGTGTA 3′) and TXT4R (5′ CGTTCGCGCCACAGTTG 3′)
   - **Amplicon Size:** 964 bp
   - **Description:** The IS1113 insertion sequence is a transposable element unique to *Xoo*. PCR amplification using these primers yields a 964 bp product that confirms the presence of the pathogen. This marker is highly specific and has been validated for use in seed health testing and quarantine diagnostics.
2. **Glycosyl Transferase Gene Marker**
   - **Primers:**
     - Forward: 5′ CCTCTATGAGTCGGGAGCTG 3′
     - Reverse: 5′ ACACCGTGATGCAATGAAGA 3′
   - **Amplicon Size:** 230 bp
   - **Description**: The glycosyl transferase gene is involved in the biosynthesis of lipopolysaccharides, a major component of the bacterial outer membrane. The targeted region is highly conserved among *Xoo* strains and produces a consistent 230 bp amplicon, making it suitable for both conventional PCR and SYBR Green–based real-time PCR detection.
3. **Xop and T3SS Gene Markers**
   - Additional molecular markers have been developed from genes encoding *Xoo* type III secretion system (T3SS) effectors such as *hrpB*, *hrpF*, and *xopQ*. These genes are functionally associated with bacterial pathogenicity and are absent or highly divergent in non-pathogenic *Xanthomonas* strains. Their detection serves as confirmatory evidence of virulent *Xoo* isolates.

6.2.4 Optimization of PCR Conditions

For reliable amplification, optimization of PCR parameters such as annealing temperature, $MgCl_2$ concentration, and primer concentration is critical. Typical PCR reaction components include the following:

- Template DNA: 50–100 ng
- Primer concentration: 0.2 µM each
- dNTPs: 0.2 mM
- Taq DNA polymerase: 1 unit
- PCR buffer (with $MgCl_2$): 1×
- Final reaction volume: 25 µL

A general thermal cycling profile includes an initial denaturation at 94 °C for 3 minutes, followed by 35 cycles of denaturation (94 °C for 30 seconds), annealing (55–60 °C for 30 seconds), and extension (72 °C for 1 minute), with a final extension at 72 °C for 5 minutes. The PCR products are resolved on a 1.5% agarose gel and visualized under UV light after ethidium bromide or SYBR Safe staining.

Amplicons of expected sizes (e.g., 964 bp or 230 bp) confirm the presence of *Xoo* in the tested sample (Figure 6.2). Absence of a band or non-specific products indicates either the absence of *Xoo* or suboptimal reaction conditions.

6.2.5 Applications of *Xoo*-Specific DNA Markers

1. **Seed health testing**: DNA markers enable early detection of *Xoo* contamination in rice seeds, supporting certification and preventing pathogen dissemination through seed movement.
2. **Epidemiological studies**: Marker-based screening allows the tracking of pathogen prevalence, diversity, and distribution across different agroclimatic zones.

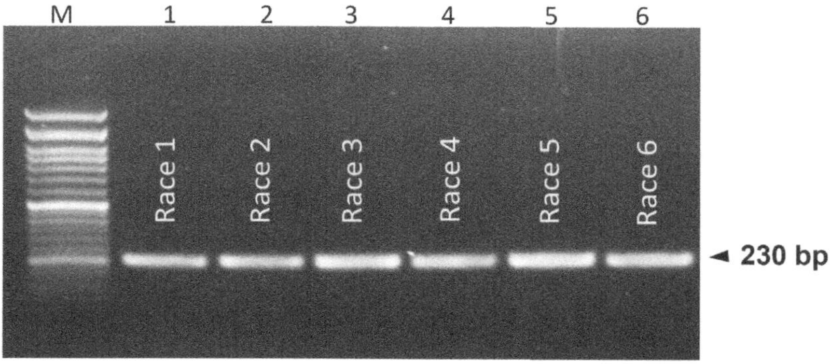

FIGURE 6.2 *Xoo*-specific amplification of glycosyl transferase gene (230 bp) across the predominant races of *Xanthomonas oryzae* pv. *oryzae*. M: Ladder of 50 bp; Lanes 1–6: Races of *Xoo* from Northern regions.

3. **Pathovar differentiation**: Specific markers help distinguish *Xoo* from other *Xanthomonas* species, such as *Xanthomonas oryzae* pv. *oryzicola*, which causes bacterial leaf streak but exhibits overlapping symptoms.
4. **Validation of resistant germplasm**: Molecular detection assists in verifying resistance screening experiments by confirming the presence or absence of *Xoo* in artificially inoculated plants.
5. **Integration with real-time PCR and LAMP assays**: The use of short, conserved target sequences allows adaptation of these markers to advanced platforms such as real-time PCR and loop-mediated isothermal amplification (LAMP), which are highly suitable for field-level diagnostics.

Marker-assisted diagnosis represents a major leap in plant pathology, enabling the detection of *Xoo* within hours instead of days. The ability to detect the pathogen even in latent infections, where visible symptoms are absent, makes this approach invaluable for quarantine and seed certification programs. Furthermore, marker-based methods provide quantitative insights into pathogen load when coupled with real-time PCR, facilitating the assessment of disease severity and progression.

These molecular tools bridge the gap between laboratory precision and field applicability, paving the way for integrated disease management strategies. As genomic databases continue to expand, new markers with improved specificity and sensitivity will further enhance the accuracy of *Xoo* diagnostics.

6.3 MULTIPLEX PCR–BASED DETECTION

Conventional PCR techniques have significantly advanced plant pathogen diagnostics; however, their capacity to differentiate among closely related bacterial species or pathovars is often limited. This limitation arises from the use of primers that may amplify conserved genomic regions shared across multiple taxa, reducing the specificity of the test. In the case of *X. oryzae* pathovars, such as those causing bacterial blight (*Xoo*) and bacterial leaf streak (*Xoc*), overlapping genetic sequences can lead to diagnostic ambiguity. Therefore, a more refined and efficient strategy is required for accurate identification.

Multiplex PCR serves as a powerful improvement over a single-target PCR assay. This technique enables the simultaneous amplification of multiple DNA targets in a single reaction mixture using several primer sets. Each primer pair is designed to amplify a unique DNA fragment corresponding to a particular pathogen, strain, or gene region, thereby allowing for concurrent detection and differentiation. In the context of *Xoo* diagnostics, multiplex PCR allows for the simultaneous identification of multiple *Xanthomonas* pathovars within a single sample, making it particularly useful for disease surveillance and mixed infection studies.

In practical terms, the diagnostic power of multiplex PCR lies in the generation of distinct amplicon sizes for different targets. These amplicons can be clearly separated and visualized on agarose gels, providing unambiguous identification. For instance, a multiplex reaction may include primers targeting *Xoo*-specific insertion sequences, virulence-related genes, or effector-associated loci, alongside internal control primers that confirm DNA quality and amplification efficiency.

A key advantage of multiplex PCR is its ability to maintain similar detection limits even when dealing with complex DNA templates containing mixtures of *Xanthomonas* strains. This indicates that the amplification efficiency of each primer set is not significantly compromised by the presence of other DNA templates. As a result, the method is highly reliable for simultaneous detection, strain differentiation, and epidemiological analysis, while saving time, reagents, and labour.

6.4 BIO-PCR AND BIO-ENRICHMENT PCR-BASED DETECTION

While conventional PCR and multiplex PCR have transformed diagnostic workflows, their effectiveness may be reduced when pathogen populations in the sample are extremely low, such as during early stages of infection or in symptomless carrier tissues. Under these conditions, even the most sensitive assays may yield false negatives due to insufficient template DNA. To address this limitation, techniques such as Bio-PCR and Bio-enrichment PCR have been developed to enhance detection sensitivity for *Xoo*.

6.4.1 Bio-PCR

Bio-PCR refers to a PCR-based technique in which plant extracts obtained from infected tissues or other biological samples are used directly as the template for amplification, without a prior enrichment step, and is widely applied in seed testing and routine diagnostics. In this approach, the standard PCR protocol is modified to include a brief, optimized sample preparation step—often involving Proteinase K treatment (e.g., ~15 minutes)—to facilitate cell lysis and release of bacterial DNA suitable for amplification.

6.4.2 Bio-Enrichment PCR

In cases where pathogen concentration is extremely low or where inhibitory compounds from plant tissues interfere with amplification, Bio-enrichment PCR offers a further refinement. The process involves incubating infected plant material in a suitable nutrient broth for an extended period, typically overnight or up to 24 hours, to allow substantial bacterial proliferation. The enrichment culture thus acts as a concentrated source of the target organism, providing abundant DNA for subsequent PCR detection.

Once the incubation period is complete, a small aliquot of the enriched broth serves as the DNA template for PCR amplification using *Xoo*-specific primers. This strategy effectively bridges the sensitivity gap observed in traditional PCR or Bio-PCR assays. It allows the detection of *Xoo* even in samples containing extremely low inoculum levels, such as asymptomatic plants, latently infected seeds, or environmental reservoirs.

Furthermore, Bio-enrichment PCR is well-suited for routine surveillance, quarantine screening, and seed certification programs. It combines the simplicity of standard PCR with a pre-enrichment step that markedly improves diagnostic reliability without requiring sophisticated instrumentation or expensive reagents.

6.4.2.1 Advantages of Bio-Enrichment PCR

- **Enhanced sensitivity:** Enables detection of *Xoo* in samples with very low bacterial concentrations.
- **Improved accuracy:** Reduces false negatives often encountered in direct PCR assays.
- **Ease of implementation:** Requires minimal equipment and can be performed in basic laboratory setups.
- **Applicability:** Ideal for seed testing, epidemiological monitoring, and latent infection diagnosis.
- **Cost-effectiveness:** Offers a robust and economical alternative to high-end molecular diagnostics.

A schematic representation of the Bio-enrichment PCR workflow for pathogen load enhancement is shown in Figure 6.3.

This approach has proven exceptionally valuable in circumstances where conventional PCR or Bio-PCR techniques fail to confirm the presence of the pathogen. Its effectiveness is particularly evident in the following situations:

- when the pathogen inoculum in the sample is extremely low,
- when visible disease symptoms are faint or indistinct,
- when symptoms are entirely absent despite a known history of disease occurrence in the field, or
- when routine diagnostic assays yield inconsistent or inconclusive results.

Such cases often create uncertainty in disease diagnosis and may result in inaccurate interpretation or misinformed management decisions. By markedly

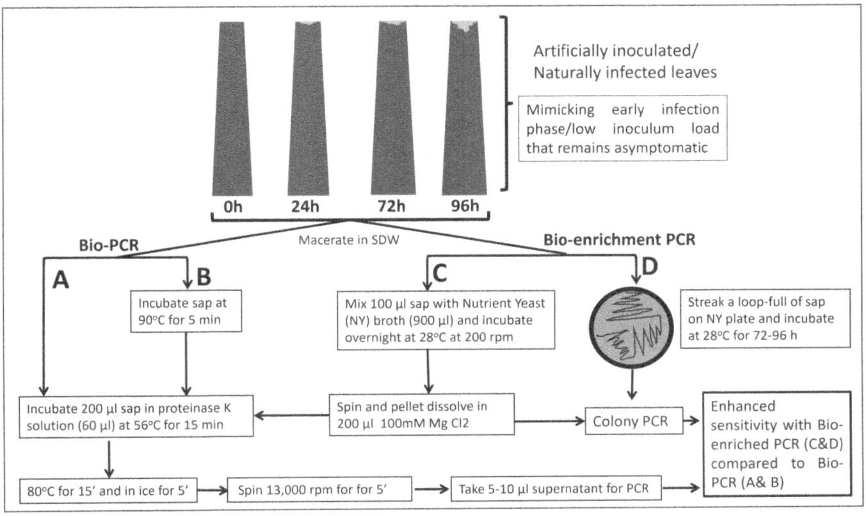

FIGURE 6.3 Workflow of Bio-PCR and Bio-enrichment PCR for enhancing the detection limit.

enhancing detection sensitivity, Bio-enrichment PCR provides a more reliable and precise diagnostic alternative. Its capability to identify *Xoo* even in low-inoculum or symptomless samples makes it indispensable for comprehensive disease surveillance, early warning systems, and timely implementation of control strategies.

6.5 REAL-TIME PCR-BASED APPROACH

Polymerase chain reaction (PCR) is among the most transformative tools in molecular plant pathology, enabling the amplification of pathogen-specific DNA sequences in a precise and reproducible manner. While conventional and multiplex PCR assays have been widely used for detecting *Xoo*, they are often limited by their qualitative nature and reduced sensitivity, particularly when pathogen concentration in plant tissues is very low. Real-time PCR has therefore emerged as a superior diagnostic platform that allows the detection, quantification, and verification of the pathogen in a single closed-tube system, minimizing the risk of contamination while providing real-time monitoring of amplification events.

The advantage of real-time PCR lies primarily in its ability to measure the accumulation of amplified DNA as the reaction proceeds. The method integrates amplification and detection, eliminating the need for post-PCR gel electrophoresis. In the context of *Xoo* detection, this feature is crucial because it allows rapid confirmation of infection status, even at early stages when visual symptoms are absent or minimal. Several studies have demonstrated that real-time PCR can detect the presence of *Xoo* DNA in leaf samples where conventional PCR or Bio-PCR fail to generate any visible bands, particularly under field conditions where the pathogen load is low.

The most widely used chemistry for real-time PCR in plant pathogen detection is based on the DNA intercalating dye SYBR Green I. The dye binds preferentially to double-stranded DNA and emits fluorescence upon excitation, which is recorded by the thermal cycler during each amplification cycle. The increase in fluorescence intensity is directly proportional to the amount of double-stranded DNA synthesized, providing a quantitative measurement of the target nucleic acid. The instrument software plots this data as an amplification curve from which the threshold cycle (Ct) value is calculated. The Ct corresponds to the cycle number at which the fluorescence signal surpasses a predefined threshold above the baseline. The lower the Ct value, the higher the initial concentration of target DNA, allowing semi-quantitative or absolute quantification of *Xoo* in infected plant material.

The choice of primers is critical in SYBR Green–based assays. Primers must be designed to amplify specific, conserved regions of the *Xoo* genome that are absent in other xanthomonads or non-pathogenic bacterial species inhabiting the phyllosphere. The *hrp* gene cluster and effector-associated loci such as avrBs3 homologues are frequently targeted because they are unique to pathogenic strains. To prevent non-specific amplification, primer pairs are typically evaluated through in silico analysis using databases such as BLAST and empirically validated by melt curve analysis. The melt curve is a post-amplification procedure that gradually increases the reaction temperature while continuously measuring fluorescence. Each amplicon exhibits a specific melting temperature (Tm) at which the double-stranded DNA dissociates.

The presence of a single sharp peak at the expected Tm indicates specificity, whereas multiple peaks signify non-specific amplification or primer-dimer formation.

Real-time PCR assays for *Xoo* require careful sampling and phenotypic screening of plant tissues prior to DNA extraction. Selection of rice leaves should be guided by detailed observation of symptom expression, including yellowing or water-soaked lesions, while also incorporating asymptomatic leaves from the same or adjacent plants for comparison. This ensures that both latent and active infections are represented in the diagnostic sample set. DNA extraction protocols must yield high-purity nucleic acids free from polysaccharides and phenolic compounds, which can inhibit the polymerase enzyme. Commercial kits or modified CTAB-based methods are commonly employed for this purpose. The purity and concentration of the DNA template are routinely verified using spectrophotometric analysis at 260/280 nm ratios before proceeding with amplification.

The thermal cycling parameters in SYBR Green assays typically include an initial denaturation step followed by 35–40 cycles of denaturation, annealing, and extension. Each step must be optimized to achieve maximal efficiency. Reaction efficiency, which ideally falls between 90% and 110%, is calculated from the slope of a standard curve generated by serially diluting known concentrations of *Xoo* genomic DNA. The coefficient of determination (R^2) indicates the linear relationship between Ct values and template quantity, with values close to 1 signifying high assay reliability. The detection limit of well-optimized real-time PCR assays for *Xoo* has been reported to reach as low as 10 fg of bacterial DNA or approximately 10^2 cfu mL^{-1}, which represents a ten- to hundred-fold improvement over conventional PCR sensitivity.

One of the major benefits of real-time PCR is its capacity for high-throughput screening. Modern instruments can process 96 or 384 reactions simultaneously, enabling large numbers of field samples to be analyzed in parallel. The closed-tube format eliminates the need to open reaction vessels after amplification, significantly reducing the risk of cross-contamination and false-positive results. Furthermore, quantitative data obtained from real-time PCR can be used for epidemiological assessments. By correlating bacterial load with disease severity or environmental conditions, researchers can gain insights into infection dynamics and pathogen population structure in different agroecological zones.

For reliable results, internal control reactions are included to account for variations in DNA extraction and amplification efficiency. These controls may target a stable host gene (such as rice actin or ubiquitin) or employ exogenous synthetic DNA fragments spiked into the samples prior to extraction. Amplification of these controls alongside the target ensures that negative results are not due to PCR inhibition. The inclusion of no-template controls (NTCs) further helps to monitor contamination or non-specific amplification.

Although SYBR Green chemistry is simple and cost-effective, it can generate false-positive signals from non-specific products. To overcome this limitation, probe-based real-time PCR assays have been developed, such as those utilizing TaqMan probes. In these systems, a fluorescent reporter and a quencher molecule are attached to opposite ends of a sequence-specific oligonucleotide that hybridizes within the target region. During amplification, the 5′ exonuclease activity of Taq polymerase cleaves the probe, separating the reporter from the quencher and producing a measurable fluorescence signal. This approach enhances specificity because fluorescence is emitted only when the probe binds and is degraded during amplification of the correct target. Although more expensive due to the synthesis of labeled probes, TaqMan-based assays are extremely reliable for confirmatory diagnostics and are suitable for multiplexing.

Multiplex real-time PCR allows simultaneous amplification of several targets in a single reaction by using different fluorescent dyes with distinct emission spectra. This configuration is useful for distinguishing among different bacterial pathovars or for including internal control genes without additional reactions. However, careful optimization is required to avoid competition among primers and probes, which can reduce efficiency or alter sensitivity. Advanced real-time PCR instruments with multiple detection channels have made such multiplexing increasingly feasible in plant pathology laboratories.

In practical applications, real-time PCR assays have been effectively used for the detection of *Xoo* from artificially inoculated and naturally infected rice plants, seed lots, irrigation water, and insect vectors. The technology has proven particularly valuable in identifying asymptomatic infections, which are often responsible for the unnoticed spread of the pathogen. By enabling rapid and sensitive quantification, real-time PCR facilitates decision-making for quarantine measures, seed certification, and integrated disease management programs.

Nevertheless, despite its precision, real-time PCR requires sophisticated equipment, trained personnel, and a controlled laboratory environment. The instrumentation involves thermal cyclers with fluorescence detection modules, and reagents must be handled with care to avoid contamination. These requirements limit its use for routine field-level screening in resource-constrained settings. The cost per reaction, though gradually decreasing, remains higher than that of simpler assays such as ELISA or lateral-flow strips. Therefore, while real-time PCR serves as the gold standard for sensitivity and quantitative accuracy, complementary serological methods are often employed for large-scale or on-site diagnostics.

In summary, real-time PCR provides a powerful platform for detecting *Xoo*, offering unparalleled sensitivity and quantitative resolution. It has overcome the main limitations of conventional PCR—namely, the need for electrophoresis and the inability to quantify pathogen load. The technique's compatibility with high-throughput workflows makes it indispensable for research laboratories and diagnostic centers. However, for rapid field-level diagnosis, there remains a need for simpler, instrument-free alternatives, which are addressed by serodiagnostic methods described in the following section.

6.6 SERODIAGNOSTICS: ENZYME-LINKED IMMUNOSORBENT ASSAY (ELISA) AND LATERAL-FLOW IMMUNOCHROMATOGRAPHIC STRIP

While molecular diagnostic platforms such as PCR and real-time PCR offer exceptional sensitivity and specificity, their dependence on advanced equipment, reagents, and skilled operators often constrains their application in routine disease monitoring. In contrast, serodiagnostic techniques exploit antigen–antibody interactions and are particularly advantageous for large-scale surveillance, field testing, and seed health certification. Immunological assays have the benefit of being simple, economical, and adaptable to various sample types, including plant tissues, seed extracts, and bacterial suspensions. For *Xoo* detection, two of the most widely utilized serodiagnostic methods are the enzyme-linked immunosorbent assay (ELISA) and the lateral-flow immunochromatographic strip (LFIA). Both techniques rely on the specific recognition between antibodies and antigens to provide qualitative or quantitative results, and both have been successfully adapted for *Xoo* diagnosis.

6.6.1 Principle and Background of Serodiagnostic Approaches

Immunodiagnostic assays are based on the specificity of antibodies that recognize unique epitopes on the surface of the target pathogen. When antibodies bind to their corresponding antigens, the resulting complexes can be visualized directly or through enzymatic or particulate labels. In the context of *Xoo* detection, antibodies raised against surface proteins, lipopolysaccharides, or extracellular polysaccharides are particularly effective because these components are pathogen-specific and abundant during infection. The use of monoclonal antibodies (mAbs) has further enhanced assay specificity and reproducibility by providing uniform binding characteristics compared with polyclonal preparations.

Among the available serological techniques, ELISA has been the most extensively adopted in plant pathology laboratories. It allows for sensitive detection of bacterial antigens through an enzyme-catalyzed colorimetric reaction. However, because ELISA involves multiple incubation and washing steps, it requires more time and laboratory infrastructure than is practical for rapid field testing. To address this limitation, the lateral-flow immunochromatographic strip assay has been developed as a simplified and portable diagnostic tool. It can be performed directly in the field with minimal training, providing visual results within minutes.

6.6.2 Development of Monoclonal Antibodies for *Xoo* Detection

The generation of monoclonal antibodies is a critical first step for both ELISA and LFIA. Typically, mice are immunized with whole-cell preparations or purified antigenic components of *Xoo* to elicit an immune response. After several booster injections, spleen cells from immunized mice are fused with immortal myeloma cells to produce hybridomas. The resulting hybridomas are screened for the production of antibodies that specifically recognize *Xoo* antigens but do not cross-react with other Xanthomonas species or common phyllosphere bacteria. Positive hybridoma clones are subcloned to ensure monoclonality and expanded to yield sufficient quantities of antibodies. These mAbs are purified from culture supernatants using affinity chromatography and then characterized for their specificity, sensitivity, and stability under varying storage conditions. Once validated, these antibodies serve as key reagents for both sandwich-ELISA and lateral-flow strip assays.

6.6.3 Sandwich-ELISA for *Xoo* Detection

The sandwich-ELISA format offers high specificity because it employs two antibodies recognizing distinct epitopes of the same antigen, one as a capture antibody and the other as a detection antibody. The method proceeds through sequential steps, as outlined next.

- **Plate coating and blocking:** Microtiter plates (usually 96-well) are coated with capture mAbs diluted in carbonate buffer (pH 9.6) at 100 μL per well. Plates are incubated at 37 °C for approximately two hours to facilitate adsorption of the antibody onto the plate surface. Following incubation, unbound antibodies are removed, and the remaining sites are blocked with a suitable blocking buffer, commonly bovine serum albumin (BSA) or skimmed milk solution. This step prevents non-specific binding during subsequent incubations.
- **Antigen binding:** Heat-killed *Xoo* cells or crude plant extracts containing bacterial cells are prepared in phosphate-buffered saline (PBS) and added to the coated wells. The antigen binds to the immobilized capture antibody during incubation at 37 °C for one hour. After incubation, plates are washed with PBS containing Tween-20 (PBST) to remove unbound material.
- **Detection antibody incubation:** A second antibody, conjugated to horseradish peroxidase (HRP), is added to each well at 100 μL per well and incubated for one hour at 37 °C. This antibody binds to a different epitope on the same antigen, forming the antibody–antigen–antibody "sandwich." Washing is again performed to remove excess conjugate.
- **Signal development and measurement:** A chromogenic substrate, typically 3,3′,5,5′-tetramethylbenzidine (TMB), is added to initiate color development. The enzyme HRP catalyzes the oxidation of TMB, producing a blue color that turns yellow upon the addition of stop solution (often 2 M sulfuric acid). Absorbance is measured at 450 nm using a microplate reader. The intensity of the color is proportional to the concentration of antigen present in the sample.
- **Assay optimization and validation:** The reliability of the sandwich-ELISA depends on careful optimization of assay parameters. The selection of the most effective capture and detection antibody pair is critical and is determined by evaluating the positive-to-negative (P/N) absorbance ratio using known positive and negative samples. A standard curve is constructed by serially diluting heat-killed *Xoo* suspensions (from 10^8 to 10^5 cfu mL^{-1}) to establish a correlation between absorbance and bacterial concentration. Cross-reactivity tests are performed with unrelated bacterial species, such as *Pseudomonas syringae*, *Burkholderia glumae*, and *Erwinia chrysanthemi*, to confirm specificity.

For validation using naturally infected or artificially inoculated plant samples, rice leaves or grains are homogenized in PBS containing Tween-20, and the extracts are clarified by centrifugation before analysis. ELISA results are interpreted based on absorbance values relative to negative controls. A sample is considered positive when its absorbance exceeds the mean negative control value by at least three standard deviations. Using this criterion, sandwich-ELISA has been shown to detect *Xoo* at concentrations as low as 1×10^4 cfu mL^{-1}, demonstrating suitability for seed health testing and phytosanitary inspections.

6.6.3.1 Performance Characteristics of Sandwich-ELISA

The ELISA method provides semi-quantitative data and can process a large number of samples simultaneously. However, the overall assay requires approximately two and a half hours to complete and depends on laboratory facilities, including microplate readers and washing systems. The sensitivity range typically spans 10^5 to 10^8 cfu mL^{-1}, which is adequate for confirming infection in symptomatic plant material but may be insufficient for latent or early-stage infections. Moreover, plant extracts occasionally contain phenolic compounds and other interfering substances that can inhibit antibody–antigen binding, leading to reduced signal intensity. Therefore, sample pre-treatment, dilution, or filtration steps are often incorporated to mitigate matrix effects. Despite these limitations, ELISA remains a cornerstone in plant pathogen diagnostics due to its cost-effectiveness, reproducibility, and capacity for large-scale testing.

6.7 LATERAL-FLOW IMMUNOCHROMATOGRAPHIC STRIP ASSAY

The lateral-flow immunochromatographic strip assay, also known as an immunostrip test, represents a significant advancement in translating laboratory-based immunodiagnostics into user-friendly field tools. It is designed for rapid, visual detection without the need for specialized instruments. The assay is based on the principle of capillary migration of liquid samples through a nitrocellulose membrane containing immobilized antibodies and visual markers.

6.7.1 Preparation of Gold Nanoparticle-Labeled Antibody Conjugates

Gold nanoparticles (GNPs) with an average diameter of 30 nm are synthesized and adjusted to pH 7.0. Monoclonal antibodies specific to *Xoo* are added to the colloidal gold solution at a final concentration of 0.1 mg mL^{-1} and incubated for approximately 50 minutes to allow passive adsorption. BSA is introduced as a stabilizing agent to prevent aggregation. The conjugates are then purified by centrifugation and resuspended in storage buffer at 4 °C until use. The strong optical properties of GNPs confer an intense red color, enabling direct visualization of assay results.

6.7.2 Assembly of the Lateral-Flow Strip

The immunochromatographic strip is composed of several functional components assembled on a plastic backing card. The sample pad, usually made of glass fiber, serves as the application site for liquid samples and is treated with surfactants and blocking agents to reduce non-specific binding. The conjugate pad, located adjacent to the sample pad, contains the GNP-labeled antibodies in a dried form. The nitrocellulose membrane, the core analytical zone of the strip, carries immobilized capture antibodies at the test line and goat anti-mouse IgG at the control line. Finally, the absorption pad facilitates the continuous flow of liquid by capillary action, ensuring that the sample migrates smoothly across the strip (Figure 6.4).

When a liquid sample, such as a plant extract or bacterial suspension, is applied to the sample pad, it rehydrates the conjugated GNP–antibody complexes, which then migrate along the membrane. If *Xoo* antigens are present, they form complexes with the GNP-labeled antibodies. These complexes are subsequently captured at

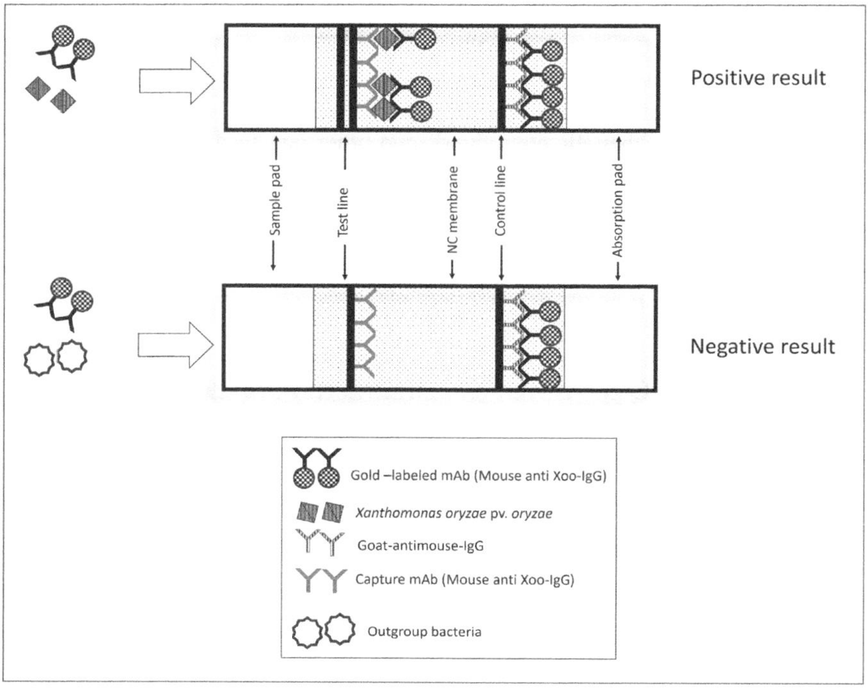

FIGURE 6.4 Schematic drawing of gold nanoparticle-labeled ELISA technique for detection of *Xoo*.

the test line by immobilized capture antibodies, generating a visible red line. The control line captures excess conjugate via anti-mouse IgG, confirming proper flow and reagent functionality. The appearance of two red lines (test and control) indicates a positive result, whereas the presence of only the control line denotes a negative result. The absence of a control line invalidates the test, signaling an operational failure.

6.7.3 Evaluation of Sensitivity and Specificity

To evaluate assay performance, serial dilutions of heat-killed *Xoo* ranging from 10^8 to 10^4 cfu mL^{-1} are tested. The lowest concentration producing a visible test line defines the cut-off value. The strip is also challenged with unrelated bacterial species to confirm specificity. Studies have demonstrated that the lateral-flow immunostrip achieves a detection limit of approximately 10^5 cfu mL^{-1} in both buffer and crude rice extract samples. Although slightly less sensitive than ELISA in absolute terms, the speed of detection (approximately 10 minutes) and complete absence of instrumentation make it particularly valuable for on-site diagnostics. Furthermore, the assay's stability at ambient temperatures and long shelf life allow for convenient storage and transportation, which are critical for field deployment.

6.7.4 Comparative Evaluation of ELISA and Lateral-Flow Immunochromatographic Strip Assays

The sandwich-ELISA and lateral-flow immunochromatographic strip each serve distinct yet complementary roles within the diagnostic spectrum for *Xoo*. ELISA remains the preferred method when quantitative or semi-quantitative data are required (Wu *et al.*, 2015b). It provides measurable absorbance values that can be correlated with pathogen concentration, making it suitable for research studies, seed health testing, and confirmatory diagnostics. However, the procedure demands controlled laboratory conditions, dedicated instruments, and technical expertise. The entire workflow, from sample preparation to absorbance reading, typically requires two to three hours. These factors limit its adaptability for immediate decision-making in the field.

The lateral-flow immunochromatographic strip, by contrast, transforms complex immunoassay chemistry into a rapid, user-friendly visual format. Its design eliminates the need for any washing, incubation, or color development steps. A simple extraction of the test material followed by application onto the sample pad is sufficient to produce results within minutes. This speed and simplicity make the strip assay ideal for preliminary screening, quarantine inspections, and field surveillance where rapid responses are essential. Moreover, the test can be conducted by personnel with minimal laboratory training, providing decentralized diagnostic capability.

From a practical perspective, both assays are not mutually exclusive but complementary. ELISA can be used for the initial calibration and validation of antibody performance, whereas the lateral-flow strip translates this established specificity into a portable diagnostic format. Together, they form an integrated immunodiagnostic toolkit that supports both laboratory and field-level surveillance of *Xoo*.

6.7.5 Integration of Molecular and Immunological Diagnostics

Although molecular methods such as PCR and real-time PCR have become the benchmark for sensitivity and specificity, integrating them with serological approaches can substantially strengthen diagnostic capacity. For instance, real-time PCR is highly suited for confirmatory detection and quantification in central laboratories, whereas immunostrip assays enable rapid pre-screening in nurseries, storage facilities, or field sites. Samples that yield ambiguous or positive results in immunostrip tests can subsequently be subjected to molecular confirmation. Such tiered diagnostic systems combine the strengths of both methodologies, ensuring rapid detection without compromising accuracy.

Another promising direction is the development of molecular–immunological hybrid assays, such as immunocapture-PCR. In this format, antibodies immobilized on microplates or magnetic beads selectively capture bacterial cells or their components from crude extracts. The bound material is then directly used as the template for PCR amplification without the need for DNA purification. This approach merges the high specificity of antibody–antigen recognition with the sensitivity of nucleic acid amplification, effectively bridging the gap between conventional serology and molecular diagnostics.

Integration also extends to sample processing and data management. With the growing emphasis on digital agriculture and precision plant health monitoring, field-deployable immunostrips can be linked to smartphone-based image analysis applications. Such tools enable automated interpretation of strip results through color intensity quantification, providing semi-quantitative information and digital record keeping for disease surveillance systems.

6.7.6 Recent Advances in Immunodiagnostics for *Xoo*

Continuous innovation in materials science, nanotechnology, and bioengineering has significantly enhanced the performance, sensitivity, and versatility of immunodiagnostic assays used for the detection of *Xoo*. Traditional immunoassays, while reliable and widely adopted, often faced limitations related to sensitivity, assay time, and scalability. Recent advances have addressed many of these constraints, enabling the development of next-generation immunodiagnostic platforms that are faster, more sensitive, and better suited for field deployment and large-scale surveillance of bacterial blight of rice.

Nanoparticle-based signal amplification has emerged as one of the most impactful innovations in immunodiagnostics. Nanoparticles composed of gold, silver, or quantum dots are increasingly used as labels in place of conventional enzyme conjugates. Gold nanoparticles remain the most widely used due to their strong optical properties, ease of surface functionalization, and compatibility with lateral flow formats. Their intense colorimetric signals allow clear visual detection even at low antigen concentrations, improving the sensitivity of immunostrips used for on-site diagnosis of *Xoo*.

Quantum dots represent a further advancement in nanoparticle-labeling technology. These semiconductor nanocrystals exhibit strong fluorescence, narrow emission spectra, and exceptional photostability. Their unique optical properties enable multiplex detection on a single diagnostic strip through color-coded emissions, allowing simultaneous identification of multiple targets or confirmation markers. For *Xoo* diagnostics, quantum dot–based immunoassays offer the potential to differentiate among closely related bacterial strains or to detect multiple rice pathogens within a single assay, thereby enhancing diagnostic efficiency and confidence.

Silver nanoparticles are also being explored for immunodiagnostic applications due to their strong plasmonic properties and signal enhancement capabilities. When integrated into immunoassays, silver-based labels can amplify detection signals, enabling the identification of low-abundance antigens that may be missed by traditional methods. The choice of nanoparticle material can be tailored to specific diagnostic requirements, balancing sensitivity, cost, and ease of detection.

Magnetic nanoparticles provide additional functional advantages beyond signal generation. These particles can be used to capture and concentrate target antigens or whole bacterial cells from complex plant extracts prior to immunodetection. By enabling rapid separation and purification steps using simple magnetic fields, magnetic nanoparticles reduce background interference and enhance assay sensitivity. In the context of *Xoo* diagnostics, this approach is particularly valuable for detecting low pathogen loads in asymptomatic tissues or environmental samples.

Microfluidic ELISA platforms represent another major advancement in immunodiagnostic technology. These systems miniaturize the conventional ELISA workflow into compact chip–based devices that operate with microliter or even nanoliter volumes of samples and reagents. Enhanced surface-to-volume ratios within microchannels improve reaction kinetics, leading to faster antigen–antibody binding and reduced assay times. Efficient fluid handling and automation further minimize operator variability and reduce the likelihood of procedural errors.

Portable microfluidic ELISA devices have demonstrated the ability to detect bacterial pathogens within 15–30 minutes, a significant improvement over traditional plate–based ELISA formats that often require several hours. For *Xoo* detection, such rapid turnaround times enable near real-time decision-making in the field. Integration of these devices with smartphone-based imaging and analysis systems allows quantitative interpretation of results without the need for bulky laboratory equipment, making them highly attractive for decentralized plant disease diagnostics.

The emergence of recombinant antibody technologies has further transformed immunodiagnostics. Recombinant antibodies and antibody fragments such as single-chain variable fragments (scFv) offer several advantages over conventional polyclonal and monoclonal antibodies. These engineered molecules exhibit improved stability, reduced batch-to-batch variability, and lower production costs. They can be expressed in bacterial or yeast expression systems at a large scale, ensuring consistent supply and performance.

ScFv fragments, due to their small size and high binding specificity, are particularly well suited for integration into biosensors, ELISA formats, and lateral flow assays. Their reduced steric hindrance allows higher surface density on sensor platforms, enhancing binding efficiency and assay sensitivity. For *Xoo* diagnostics, recombinant antibody–based assays provide reliable performance and facilitate standardization across laboratories and field applications.

In parallel with antibody engineering, aptamer-based detection strategies have gained increasing attention. Aptamers are short single-stranded nucleic acids that fold into specific three-dimensional structures capable of binding target molecules with high affinity and specificity. Unlike antibodies, aptamers can be chemically synthesized with high reproducibility and modified with functional groups for easy integration into biosensors. Their thermal stability and long shelf life make them particularly suitable for use in diverse environmental conditions.

Aptamer-based immunodiagnostic alternatives offer several advantages, including reduced dependence on animal-based antibody production and greater flexibility in target selection. For *Xoo*, aptamers can be designed to recognize specific surface proteins or virulence-associated molecules, providing highly specific detection. Their compatibility with electrochemical, optical, and colorimetric readouts positions aptamers as key components of next-generation biosensing platforms.

The integration of nanomaterials, microfluidics, recombinant antibodies, and aptamers is driving the evolution of immunodiagnostics toward fully integrated biosensor systems. These systems combine sample preparation, target recognition, signal amplification, and data readout within a single compact device. Such integration reduces assay complexity, shortens diagnostic timelines, and enhances usability for non-specialized operators.

Despite these advances, challenges remain in translating laboratory innovations into widely adopted field diagnostics. Issues related to cost, robustness, and large-scale manufacturing must be addressed to ensure accessibility, particularly in resource-limited settings. Nevertheless, the rapid pace of technological development suggests that these barriers are progressively being overcome.

In conclusion, recent advances in immunodiagnostics for *Xoo* have significantly expanded the capabilities of disease detection and surveillance. Nanoparticle-enhanced assays, microfluidic ELISA platforms, recombinant antibodies, and aptamer-based detection systems collectively represent a new generation of immunodiagnostic tools. By retaining the simplicity and reliability of traditional immunoassays while incorporating modern technological innovations, these approaches are poised to play a central role in the early detection and effective management of bacterial blight of rice.

6.8 AI-BASED TECHNIQUES

The increasing availability of high-resolution imaging systems and advances in computational algorithms have made artificial intelligence an increasingly powerful tool for disease detection in crop plants. For the diagnosis of bacterial blight caused by *Xoo*, the application of artificial intelligence techniques has shown great promise, particularly in the context of early-stage recognition and large-scale disease surveillance. These techniques integrate image processing, deep learning, and pattern recognition to detect disease symptoms on rice leaves with remarkable precision. By replacing subjective visual assessment with automated digital analysis, artificial intelligence enables faster and more consistent identification of infection in both field and controlled environments.

Artificial intelligence–based systems for bacterial blight detection depend primarily on the quality and diversity of image data. The process begins with the creation of a comprehensive image database containing digital photographs of both healthy and infected rice leaves. Such a database should include images representing different varieties, growth stages, and degrees of infection to ensure robustness of the model. The images are captured under natural daylight conditions on sunny days, using a camera with at least 13 megapixels of resolution. The flash function is avoided to prevent unwanted glare and reflection that may distort texture and color information. For meaningful statistical representation, approximately five hundred to six hundred high-resolution images, each of about 1,333 by 1,000 pixels, are typically required.

To ensure the presence of confirmed infection, diseased samples are generated by artificial inoculation of rice leaves using the leaf tip clipping technique, which introduces *Xoo* suspension at the tip of fully expanded leaves. This method produces uniform lesions that replicate natural infection. After inoculation, the plants are maintained under field conditions and monitored regularly. Image collection is carried out at seven-day intervals to document disease progression from the onset of infection to the stage of complete blighting. The first image set is taken at the seedling stage, approximately 30 days after transplanting. At this time, healthy leaves appear evenly green and glossy, whereas infected leaves begin to show minor yellow discoloration at the tip within seven days of inoculation. As infection advances, the discoloration progresses along the veins and eventually leads to drying and blighting of the leaf tissue. By the panicle initiation stage, lesions become extensive and easily visible, signifying a susceptible reaction.

To improve image quality and facilitate accurate disease recognition, care must be taken during photography. The leaves are placed against a plain black or monochrome background to minimize background noise and eliminate interference from soil, sky, or other vegetation. Shadows and reflections are avoided to maintain uniform illumination. These precautions ensure that the captured images represent true disease symptoms and not artifacts from external factors. Each image is labeled according to its health status and stage of infection, which forms the basis for subsequent model training and validation.

Once the image database is established, the next step involves processing and analyzing the images using computational techniques that extract relevant features from the leaf surfaces. The processing pipeline generally includes several stages: image acquisition, preprocessing, segmentation, feature extraction, and classification. Preprocessing involves adjusting brightness, contrast, and color balance to normalize all images and ensure consistency. Noise is removed using filtering techniques, and images are resized to a uniform dimension to facilitate efficient processing. Segmentation follows, which separates the region of interest (the leaf area) from the background. Thresholding and clustering methods such as k-means clustering are often employed for this purpose.

After segmentation, quantitative information is extracted from the images. This step, known as feature extraction, captures essential characteristics that distinguish diseased leaves from healthy ones. Multiresolution analysis using the Discrete Wavelet Transform plays a central role in this process. The Discrete Wavelet Transform decomposes each image into multiple frequency sub-bands that represent variations in texture and color at different spatial scales. Four families of wavelets are commonly applied in such analyses: Daubechies, Symlet, Biorthogonal, and Coiflet. Each family provides a unique mathematical structure for representing the image information. In the context of bacterial blight detection, the Daubechies family has demonstrated particularly strong performance due to its ability to capture subtle variations in leaf texture and lesion boundaries with high precision.

Feature extraction produces a large set of coefficients representing different aspects of the image. However, the high dimensionality of these data can increase computational complexity and reduce processing efficiency. To overcome this limitation, Principal Component Analysis is applied to reduce the dimensionality of the feature set while retaining the most significant information. Principal Component Analysis transforms the original correlated variables into a smaller set of uncorrelated components, which simplifies the data and enhances the performance of subsequent classification models. The combination of Discrete Wavelet Transform and Principal Component Analysis thus forms a powerful feature extraction framework capable of identifying fine-grained patterns in leaf images.

The classification stage is implemented using Deep Neural Networks, which are advanced machine learning architectures inspired by the human brain. These networks consist of multiple layers of interconnected neurons that progressively learn abstract representations from input data. In this application, three widely used deep neural network architectures, AlexNet, ResNet101, and Inception V3, are employed for the classification of rice leaf images into healthy and diseased categories. Each network architecture has a distinct design and depth, influencing its capacity to capture different features from the input data.

AlexNet, with its eight-layer structure, was one of the earliest deep networks that revolutionized image recognition by introducing rectified linear activation functions and dropout layers to prevent overfitting. ResNet101, on the other hand, incorporates residual learning through skip connections that allow the network to train efficiently even at great depth. Inception V3 employs a more complex structure with multiple convolutional kernels operating in parallel, enabling it to capture both fine and coarse details in an image.

Comparative evaluation of these architectures revealed that Inception V3 provided the highest accuracy for detecting bacterial blight symptoms on rice leaves. The system achieved optimal performance when feature extraction was performed at decomposition level one using the approximation coefficient sub-band generated by the Discrete Wavelet Transform. Under these conditions, the model reached an impressive accuracy of 91%, outperforming the other two architectures. Among the tested wavelet families, the Daubechies wavelet produced the best classification results, confirming its suitability for texture-based image analysis in plant pathology.

For validation and benchmarking, the system was also tested using four publicly available datasets from the PlantVillage database, which included images of apple, corn, grape, and tomato leaves. The performance consistency across these datasets demonstrated the robustness of the artificial intelligence framework for plant disease diagnosis. When the model was applied to the Rice Plant Disease dataset, it successfully distinguished bacterial blight symptoms from healthy tissue with near-perfect precision.

The complete artificial intelligence–based workflow therefore encompasses image acquisition, preprocessing, segmentation, feature extraction through multiresolution analysis, dimensionality reduction using Principal Component Analysis, and final classification by Deep Neural Networks. Each stage contributes to enhancing the accuracy and efficiency of the overall system. The integration of wavelet analysis with deep learning allows the model to detect intricate spatial variations in color and texture that are often imperceptible to the human eye.

The use of artificial intelligence offers several advantages over conventional diagnostic methods. It eliminates the need for destructive sampling and laboratory testing, allowing in-field disease assessment through simple image capture. The process is rapid, objective, and easily scalable to large agricultural landscapes. Once trained, the system can operate autonomously, providing farmers and researchers with instant feedback on disease presence and severity. Additionally, it can be integrated with mobile applications or unmanned aerial vehicles for real-time surveillance and mapping of disease incidence.

The study demonstrated that such systems are capable of processing large-scale datasets that are often challenging to analyze using traditional machine learning methods. By leveraging deep learning architectures, the model effectively learned discriminative features that corresponded to the characteristic lesion patterns caused by *Xoo* infection. As reported by Kirti *et al.* (2022) and Azizi and Lau (2022), this integration of artificial intelligence and image analysis provides a highly reliable approach for disease recognition, achieving superior accuracy compared with manually engineered feature extraction techniques.

In conclusion, artificial intelligence–based techniques represent a significant advancement in the field of plant disease diagnostics. By combining high-quality image data with advanced computational models, it is now possible to achieve early detection of bacterial blight with exceptional precision. Such systems not only aid researchers in understanding disease epidemiology but also empower farmers with accessible digital tools for timely disease management. With continuous improvement in imaging devices and computational power, the integration of artificial intelligence into agricultural diagnostics will continue to expand, making precision agriculture more efficient and sustainable.

6.9 EMERGING MOLECULAR APPROACHES COMPLEMENTING SERODIAGNOSTICS

Advances in molecular diagnostics continue to significantly augment the detection landscape for *Xoo*, providing powerful tools that complement conventional serodiagnostic methods. While immunological assays such as ELISA and lateral flow immunostrips remain central to routine screening and field-level detection, molecular approaches offer enhanced sensitivity, specificity, and quantitative capability. Together, these methodologies form a complementary diagnostic continuum, allowing flexible deployment across laboratory, semi-field, and field settings.

Among molecular techniques, digital PCR represents a major refinement of conventional real-time PCR technology. Digital PCR enables absolute quantification of target nucleic acids by partitioning the reaction mixture into thousands to millions of nanoliter-scale droplets or micro-chambers, each acting as an independent amplification unit. Following amplification, droplets are scored as positive or negative, and the absolute copy number of the target DNA is calculated using statistical models. This partitioning significantly reduces the impact of PCR inhibitors commonly present in plant tissues and environmental samples, thereby enhancing analytical precision and reproducibility. For *Xoo* detection, digital PCR is particularly valuable for identifying extremely low pathogen loads in asymptomatic plants, seed material, or early infection stages where serological assays may lack sufficient sensitivity.

The capacity of digital PCR to detect even a single copy of target DNA has important implications for epidemiological studies and resistance screening. Absolute quantification allows researchers to precisely compare pathogen populations across samples, locations, and time points. This is especially useful for studying pathogen dynamics in resistant and susceptible rice genotypes, evaluating the effectiveness of management interventions, and tracking the early establishment of *Xoo* in new environments. Despite these advantages, the relatively high cost of digital PCR platforms and reagents currently limits their widespread deployment beyond research laboratories and specialized diagnostic facilities.

Parallel to advances in digital PCR, the development of portable real-time PCR instruments has brought molecular diagnostics closer to practical field application. Compact devices powered by rechargeable batteries and controlled through smartphone or tablet interfaces are increasingly available. These systems retain the sensitivity and specificity of laboratory-based real-time PCR while offering enhanced mobility and ease of use. For *Xoo* detection, portable real-time PCR platforms enable on-site confirmation of suspected infections, reducing diagnostic turnaround time and facilitating timely management decisions. Such tools are particularly valuable in remote or resource-limited areas where access to centralized laboratories is constrained.

Isothermal nucleic acid amplification technologies have emerged as transformative approaches for field-adapted molecular diagnostics. Loop-mediated isothermal amplification (LAMP) amplifies target DNA at a constant temperature using a set of specially designed primers and a strand-displacing DNA polymerase. The reaction produces large quantities of amplified DNA within a short time, often less than one hour. Amplification can be detected through turbidity, colorimetric indicators, fluorescence, or lateral flow formats, allowing flexible readout options. For *Xoo* diagnostics, LAMP assays have demonstrated high sensitivity and specificity comparable to real-time PCR, while offering operational simplicity.

Recombinase polymerase amplification (RPA) represents another promising isothermal technique that operates at relatively low and constant temperatures. RPA reactions typically proceed rapidly, often yielding detectable amplification within 15–30 minutes. The low temperature requirement allows amplification to be carried out using simple heating devices or even body heat, further enhancing suitability for field deployment. Like LAMP, RPA products can be visualized using lateral flow strips, fluorescence readers, or colorimetric indicators. These features make RPA particularly attractive for rapid on-site detection of *Xoo* in field conditions.

The integration of isothermal amplification methods with lateral flow readouts exemplifies the trend toward hybrid diagnostic platforms that combine molecular sensitivity with visual simplicity. In such systems, amplified nucleic acids are detected using immunochromatographic strips similar in format to antibody-based immunostrips. The appearance of visible bands provides immediate confirmation of pathogen presence without the need for specialized equipment. This convergence of molecular and serological principles bridges the gap between laboratory-grade diagnostics and farmer-friendly tools, enabling broader access to reliable disease detection.

Emerging molecular approaches also benefit from innovations in sample preparation and nucleic acid extraction. Simplified extraction methods compatible with isothermal amplification reduce assay complexity and turnaround time. Crude extraction buffers, paper-based sample processing, and cartridge-integrated workflows are increasingly being explored to minimize handling steps and reduce the risk of contamination. For *Xoo* diagnostics, streamlined sample preparation is critical for effective field use, particularly when assays are performed by extension personnel or farmers with limited technical training.

Another important dimension of emerging molecular diagnostics is multiplexing capability. Molecular assays can be designed to simultaneously detect multiple pathogens or distinguish among closely related strains. This is particularly relevant in rice ecosystems where multiple bacterial, fungal, and viral pathogens may co-occur and produce overlapping symptoms. Complementing serodiagnostics with multiplex molecular assays enhances diagnostic confidence and supports more precise disease management strategies.

Despite their advantages, molecular approaches are most effective when deployed alongside serodiagnostic tools rather than as standalone solutions. Serodiagnostics offer rapid, low-cost screening suitable for large-scale surveys, while molecular methods provide confirmatory and quantitative information. Integrating these approaches within a tiered diagnostic framework allows efficient allocation of resources and maximizes diagnostic coverage across scales. For example, immunostrips may be used for initial field screening, followed by isothermal amplification or real-time PCR for confirmation of critical samples.

Looking ahead, continued miniaturization, cost reduction, and integration of molecular diagnostics are expected to further enhance their complementarity with serodiagnostics. The incorporation of molecular assays into portable, user-friendly devices with simple visual or digital outputs will expand their adoption beyond specialized laboratories. As these technologies mature, they will play an increasingly important role in early detection, surveillance, and management of bacterial blight of rice.

In summary, emerging molecular approaches such as digital PCR, portable real-time PCR, LAMP, and RPA significantly strengthen the diagnostic toolbox for *Xoo*. By complementing established serodiagnostic methods, these technologies enable sensitive, specific, and adaptable detection across diverse operational contexts. Their integration into comprehensive diagnostic frameworks will be central to improving disease surveillance, supporting timely management decisions, and ultimately reducing the impact of bacterial blight on global rice production.

6.10 FUTURE PROSPECTS FOR INTEGRATED DIAGNOSTIC SYSTEMS

The convergence of nanotechnology, microfluidics, and molecular biology is steadily driving the evolution of plant pathogen diagnostics toward compact, rapid, and highly sensitive formats. In recent years, advances in materials science, sensor engineering, and bio-recognition chemistry have collectively transformed how plant diseases are detected and monitored. For *Xoo*, the causal agent of bacterial blight of rice, these developments are particularly significant, given the need for early detection, rapid response, and large-scale surveillance across diverse rice-growing environments. The future of *Xoo* diagnostics lies in the development of integrated diagnostic systems that seamlessly combine sensitivity, specificity, portability, and data connectivity.

Integrated diagnostic platforms aim to consolidate multiple recognition and signal transduction modalities into a single device or workflow. Rather than relying on a single principle of detection, such systems exploit complementary mechanisms to improve accuracy and robustness. Biosensors that integrate immunochemical recognition with electrochemical, optical, or piezoelectric signal generation are already under active exploration. In these systems, antibody–antigen binding events are converted into measurable electrical or optical signals that can be processed electronically for quantitative analysis. Such integration reduces assay time, minimizes reagent consumption, and enhances detection sensitivity, making diagnostics more suitable for both laboratory and field applications.

Nanotechnology plays a central role in shaping next-generation diagnostic systems. Nanomaterials such as gold nanoparticles, quantum dots, carbon nanotubes, and graphene-based composites offer unique electrical, optical, and surface properties that enhance signal amplification and detection limits. In immunodiagnostic assays, nanoparticles can be functionalized with antibodies or antigens to increase binding efficiency and improve visual or electronic readouts. Similarly, nanostructured electrodes in electrochemical biosensors provide increased surface area and improved electron transfer, enabling the detection of extremely low concentrations of *Xoo* cells or biomolecules.

Microfluidic technologies further contribute to the miniaturization and integration of diagnostic systems. Lab-on-chip platforms allow precise manipulation of minute volumes of samples and reagents within enclosed channels, reducing assay time and contamination risks. For *Xoo* detection, microfluidic devices can integrate sample preparation, reaction, and detection steps into a single compact unit. This is particularly advantageous for field diagnostics, where access to laboratory infrastructure is limited. The ability to process multiple samples simultaneously also supports high-throughput screening during disease surveys and epidemiological studies.

Artificial intelligence and machine learning are increasingly being incorporated into diagnostic pipelines to automate data interpretation and decision-making. Digital imaging systems coupled with trained algorithms can objectively interpret color intensity, band thickness, or fluorescence output from ELISA plates, lateral flow immunostrips, or biosensor surfaces with high precision. Such approaches reduce operator-dependent variability and improve diagnostic reliability, especially when assays are deployed by non-specialized users. Over time, machine learning models can be continuously refined using large datasets, further enhancing their predictive accuracy and adaptability to new diagnostic formats.

The integration of cloud-based data storage and connectivity represents another major advancement in future diagnostic frameworks. Diagnostic results generated in the field can be uploaded in real time to centralized databases, enabling regional and national monitoring of bacterial blight incidence. Such systems support early warning networks, spatial disease mapping, and trend analysis, thereby strengthening plant health surveillance programs. For *Xoo*, which exhibits significant genetic and virulence diversity across regions, centralized data platforms can also facilitate the correlation of diagnostic outcomes with pathogen population dynamics and resistance deployment strategies.

In the context of sustainable agriculture, future diagnostic assays must align with environmentally responsible practices. The development of biodegradable substrates for test strips, nontoxic chromogens, and recyclable or low-waste cartridges is gaining increasing attention. Reducing reliance on hazardous chemicals and minimizing diagnostic waste are important considerations, particularly when assays are used at large scale. Energy-efficient devices, including those powered by batteries or solar panels, further enhance the sustainability and practicality of field-deployable diagnostic systems.

Capacity building and user engagement are equally critical to the success of integrated diagnostic technologies. Training programs that promote the correct use and interpretation of immunodiagnostic and molecular tools among farmers, extension workers, and field technicians are essential for widespread adoption. User-friendly designs, clear visual outputs, and minimal sample handling requirements increase the likelihood that diagnostic tools will be effectively utilized under real-world conditions. Participatory approaches that involve end users in the evaluation and refinement of diagnostic devices can further improve usability and acceptance.

Despite the rapid progress in integrated diagnostics, molecular tools such as real-time PCR remain central to the detection and identification of *Xoo*. Real-time PCR continues to be the most sensitive and quantitative molecular method available, capable of detecting minimal pathogen loads even in asymptomatic tissues. Its high specificity and reproducibility make it indispensable for confirmatory diagnostics, research applications, and detailed epidemiological investigations. However, the dependence of real-time PCR on sophisticated laboratory infrastructure, stable power supply, and skilled personnel limits its routine use in field settings.

Serodiagnostic methods provide practical and complementary alternatives to molecular assays. Techniques such as sandwich-ELISA and lateral flow immunochromatographic strips offer robust, cost-effective, and accessible options for *Xoo* detection. ELISA-based assays enable high-throughput quantitative screening of large sample numbers in laboratory environments, supporting surveillance and resistance screening programs. In contrast, lateral flow strips provide immediate qualitative results, making them ideal for rapid field diagnosis and on-site decision-making. Their simplicity, affordability, and minimal equipment requirements have contributed to their widespread adoption in plant disease diagnostics.

Future diagnostic frameworks for *Xoo* are expected to integrate these established serological and molecular assays with emerging biosensing, digital imaging, and portable amplification technologies. Isothermal nucleic acid amplification methods, combined with simple visual or electronic readouts, hold promise for bridging the gap between laboratory-grade sensitivity and field-level usability. When coupled with immunodiagnostics and biosensors, such hybrid systems can provide layered confirmation, enhancing confidence in diagnostic outcomes.

The ultimate goal of integrated diagnostic systems is to deliver platforms that are sensitive, specific, rapid, inexpensive, and adaptable to diverse environmental and socio-economic contexts. For bacterial blight of rice, timely and accurate diagnosis underpins effective disease management, from targeted chemical interventions to the deployment of resistant varieties. Integrated diagnostics also support proactive disease monitoring, enabling early intervention before widespread epidemics occur.

In conclusion, the future prospects for integrated diagnostic systems for *Xoo* are shaped by advances in nanotechnology, microfluidics, artificial intelligence, and sustainable design. By combining established diagnostic methods with innovative sensing and data management technologies, next-generation platforms can transform how bacterial blight is detected and managed. Such systems will play a crucial role in supporting sustainable rice production and strengthening global efforts to mitigate the impact of one of the most destructive diseases affecting this vital crop.

7 Virulence Screening of *Xoo* Races and Pathotyping

7.1 RACE SCENARIO

Several races of *Xoo* have been reported both nationally and globally, and these races differ in their virulence patterns, as established through early and contemporary studies (Mondal *et al.*, 2014; Yugander *et al.*, 2017). The identification of these races forms the foundation of understanding pathogen population structure and its impact on resistance deployment strategies. Over the past two decades, extensive pathotyping efforts have revealed substantial spatial variation in race prevalence, shaped by host genotype distribution, agroecological factors, cropping intensity, vectorless dissemination, monsoon patterns, and local adaptation of pathogen lineages. A national framework for race surveillance has become increasingly important, as localized surveys are insufficient to capture the dynamic shifts in *Xoo* populations.

To assess virulence, researchers commonly use standard sets of near-isogenic lines (NILs) carrying individual resistance genes, along with pyramided lines developed at IRRI, Philippines (Table 7.1). These genetic differentials provide a stable platform for comparing isolates across studies and regions. Data from IRRI's global screening network, particularly the International Rice Bacterial Blight Nursery (IRBBN), have been instrumental in standardizing virulence characterization across diverse rice-growing regions. However, despite these resources, India still lacks a fully synchronized national program that integrates regional, seasonal, and molecular-level surveillance, thereby limiting a comprehensive understanding of race distribution. It is now widely recognized that *Xoo* races are not only geographically variable but also temporally unstable, with newly emerging races often showing higher aggressiveness. These evolutionary changes create a pressing need for systematic, long-term monitoring of race patterns to guide resistance gene deployment, breeding programs, and disease forecasting systems.

Race identification is further complicated by the fact that many *Xoo* populations exhibit intermediate or atypical reactions that do not fit neatly into classical race frameworks. Studies in South and Southeast Asia suggest that race structures are evolving at accelerating rates due to the intensive introduction of new rice cultivars carrying single major resistance genes. Such directional selection pressures lead to the rapid emergence of virulence variants capable of overcoming widely deployed *Xa* genes. Consequently, understanding the race scenario today requires combining classical phenotypic assays with molecular markers and TAL effector profiling, which

DOI: 10.1201/9781003714590-7

TABLE 7.1
NILs and Pyramided Rice Lines

NILs	*R* Genes	Pyramided Lines	*R* Gene Combinations
IRBB1	*Xa1*	IRBB50	*Xa4+ xa5*
IRBB2	*Xa2*	IRBB51	*Xa4+ xa13*
IRBB3	*Xa3*	IRBB52	*Xa4+Xa21*
IRBB4	*Xa4*	IRBB54	*xa5+Xa21*
IRBB5	*xa5*	IRBB55	*xa13+Xa21*
IRBB7	*Xa7*	IRBB56	*Xa4+ xa5+ xa13*
IRBB8	*xa8*	IRBB57	*Xa4+ xa5+ Xa21*
IRBB10	*Xa10*	IRBB59	*xa5+ xa13+ Xa21*
IRBB11	*Xa11*	IRBB60	*Xa4+ xa5+ xa13+ Xa21*
IRBB13	*xa13*	IRBB61	*Xa4+ xa5+ Xa7*
IRBB14	*Xa14*	IRBB62	*Xa4+Xa7+Xa21*
IRBB21	*Xa 21*	IRBB63	*xa5+ Xa7+ xa13*
		IRBB64	*Xa4+ xa5+ Xa7 + Xa21*
		IRBB65	*Xa4+ Xa7+ xa13 + Xa21*
		IRBB66	*Xa4+ xa5+ Xa7+Xa13+ Xa21*

together provide deeper insight into the evolutionary basis of race shifts. The evolving race landscape underscores the need for continued, expanded, and coordinated surveillance across all rice-growing belts in India to maintain an updated inventory of virulence patterns and emerging threats.

Mondal *et al.* (2014) conducted an extensive race screening program across northern and eastern India, revealing clear regional structuring of *Xoo* populations. Race 4 and Race 6 were predominant in states such as Punjab, Haryana, Himachal Pradesh, Uttarakhand, Uttar Pradesh, and Rajasthan, reflecting possibly long-term adaptation to dominant rice cultivars grown in those regions (Table 7.2). In contrast, Race 2 and Race 3 were more common in the eastern states, including West Bengal, Bihar, Jharkhand, Assam, Nagaland, and Meghalaya. Race 1 and Race 5 occurred sporadically across both ends of the country, suggesting a wider but lower abundance distribution. These findings highlight the importance of spatially resolved surveillance in characterizing the diversity and movement of *Xoo* races.

7.2 ARTIFICIAL INOCULATION FOR VIRULENCE SCREENING

Artificial inoculation remains the cornerstone of virulence screening for *Xoo*, despite the availability of rapid molecular detection tools. Classical inoculation assays provide direct evidence of pathogenicity, lesion development, and host–pathogen interactions. They offer a robust means for comparing isolates across studies, enabling uniform scoring standards to be applied to race determination. Artificial inoculation

TABLE 7.2
Reaction of Six Races from North and Eastern India on Rice NILs

		Disease Reaction (Mean of 15 Leaves)<5=R, 5-<10 = MR, 10–15=MS, >15=S					
NILS	***R* Genes**	**Race 1 (Meghalaya, ICAR Umain)**	**Race 2 (Assam, Nangpoh)**	**Race 3 (HP, Kangra 2)**	**Race 4 (Mathura, UP)**	**Race 5 (WB-Kashipur)**	**Race 6 (Rajasthan, Banswara)**
IRBB1	*Xa1*	MS	MR	MR	MS	MS	S
		(10.33)	(5.5)	(8.33)	(10.1)	(11.5)	(15.33)
IRBB2	*Xa2*	S	S	S	S	S	S
IRBB3	*Xa3*	MS	S	MR	S	S	S
		(11.16)	(16.16)	(8.33)	(15.2)	(28.34)	(29)
IRBB4	*Xa4*	S	R	MR	MS	MS	S
		(23.7)	(4)	(9.8)	(12.1)	(14.33)	(16)
IRBB5	*xa5*	MR	R	MR	MS	MR	MS
		(9.8)	(4.7)	(8.33)	(13.4)	(8.33)	(11.33)
IRBB7	*Xa7*	MR	R	MR	MS	MS	S
		(6)	(5.5)	(7.7)	(12)	(10.33)	(28.33)
IRBB8	*xa8*	R	MR	MR	MS	MR	S
		(4)	(7.8)	(6)	(11.6)	(6.67)	(11.33)
IRBB10	*Xa10*	MR	MS	MS/S	MS	MS	S
		(6.33)	(10.5)	(11.0)	(10.8)	(10.33)	(10.7)
IRBB11	*Xa11*	MR	MR	MR	MS	MS	S
		(10)	(6)	(7)	(11.2)	(11.3)	(17)
IRBB13	*xa13*	R	R	R	MR	R	R
		(3)	(1.7)	(3.7)	(9.8)	(2.3)	(4)
IRBB14	*Xa14*	S	R/MR	S	S	MS	S
		(15.1)	(6)	(15.5)	(18.4)	(12)	(2.8)
IRBB21	*Xa21*	R	R	R	MR	MR	MR
		(4.7)	(4.5)	(2.5)	(9.27)	(6.5)	(8.7)
IRBB50	*Xa4 + xa5*	MR	R	R	MS	R	MS
		(6.33)	(4.33)	(4.3)	(13.7)	(4.7)	(10.66)
IRBB51	*Xa4 +xa13*	R	R	R	MR	R	R
		(3.33)	(3.33)	(3)	(8.5)	(2.67)	(4.8)
IRBB52	*Xa4+Xa21*	MR	MR	MR	MR	MR	MR
		(5.5)	(5.33)	(5.8)	(9.9)	(5.5)	(6.7)
IRBB53	*xa5+xa13*	R	R	R	MR	R	MR
		(2)	(1)	(4)	(9.6)	(D	(8.7)
IRBB54	*xa5 + Xa21*	MR	R	MR	MR	MR	MR
		(6.33)	(3.7)	(7.2)	(9.5)	(7.3)	(8.2)
IRBB55	*xa13 + Xa21*	R	MR	R	MR	R	R
		(4.5)	(9.16)	(1.33)	(8.5)	(4)	(4.5)

(Continued)

TABLE 7.2 ***(Continued)***
Reaction of Six Races from North and Eastern India on Rice NILS

		Disease Reaction (Mean of 15 Leaves)<5=R, 5-<10 = MR, 10–15=MS, >15=S					
NILS	***R* Genes**	**Race 1 (Meghalaya, ICAR Umain)**	**Race 2 (Assam, Nangpoh)**	**Race 3 (HP, Kangra 2)**	**Race 4 (Mathura, UP)**	**Race 5 (WB-Kashipur)**	**Race 6 (Rajasthan, Banswara)**
IRBB56	*Xa4+xa5+*	R	R	R	MR	R	R
	xa13	(1.3)	(3.3)	(5)	(8.7)	(4.3)	(4.8)
IRBB57	*Xa4+xa5+*	MR	MR	R	MR	MR	MR
	Xa21	(5.5)	(6.33)	(4.3)	(6.1)	(6.3)	(6.7)
IRBB58	*Xa4 +xa13 +*	R	R	R	MR	R	R
	Xa21	(4.16)	(2.16)	(1)	(9.4)	(4)	(2.7)
IRBB59	*xa5 + xa13 +*	R	R	R	MR	R	R
	Xa21	(2.16)	(1.5)	(D	(8.6)	(D	(2.0)
IRBB60	*Xa4+xa5+xa13*	R	R	R	MR	MR	MR
	Xa21	(2.16)	(3.5)	(4)	(8.3)	(6.5)	(7.33)
TN1	*Xa14*	S	MS	S	S	MS	S
		(15.6)	(11)	(16.4)	(15.4)	(13)	(25.6)
Ajaya	-	S	R	MR	MS	MR	MS
		(18.3)	(4)	(5.3)	(12.5)	(5.7)	(13.7)
DV85	*Xa7*	MS	MS	MS	MS	MS	S
		(11.7)	(14)	(10)	(10.7)	(10.8)	(28)

ensures controlled infection pressure under optimized environmental conditions, which is essential for differentiating subtle phenotypic responses among NILs and pyramided lines. Given that natural infection is influenced by fluctuating environmental variables, artificial inoculation eliminates these confounding factors, thereby producing more reproducible and comparable data.

The leaf tip clipping method described by Kauffman *et al.* (1973) remains widely used due to its simplicity, reliability, and ability to produce consistent lesion development. This method enables researchers to quantify lesion progression accurately, as it bypasses the natural barriers posed by the leaf cuticle and directly introduces the pathogen into exposed vascular tissues. Proper execution of clipping, inoculum concentration, plant growth stage, and environmental conditions are essential for generating interpretable results. In many breeding and pathology programs, artificial inoculation is conducted in nurseries under high humidity to ensure uniform infection, but controlled glasshouse conditions are increasingly preferred for precision studies.

Artificial inoculation is also critical for validating the virulence phenotype of newly collected *Xoo* isolates prior to genome sequencing or molecular characterization. Without accurate virulence profiling, molecular data alone cannot reliably define pathotypes. Thus, the combination of artificial inoculation and lesion scoring forms the empirical foundation upon which race classification, resistance gene identification, and virulence evolution studies depend.

7.3 INOCULUM PREPARATION

Inoculum preparation begins with selecting a single-colony pure culture to ensure that the inoculating suspension represents a genetically homogeneous population. The culture is grown on YGCA medium (Yeast extract 10 g, Glucose 10 g, $CaCO_3$ 20 g, Agar 20 g, Distilled water to 1 L; pH 7.0) for 48 hours at 28 °C, conditions that optimize bacterial growth and yield high cell density suitable for inoculation. The bacterial growth is gently scraped using a sterile glass slide and suspended in sterile distilled water to avoid physical disruption that may affect cell viability. The optical density of the suspension is adjusted to an OD600 of 1.0, corresponding to approximately 10^9 cfu/mL, which has been empirically validated to produce reliable infection without overwhelming plant tissues.

Freshly prepared inoculum is essential for accurate virulence testing. Older suspensions may experience physiological changes, reduced motility, or uneven density due to cell aggregation, leading to inconsistent infection severity. Care must also be taken to maintain sterility throughout inoculum preparation, as contaminants can alter lesion appearance or interfere with accurate race identification. Standardizing inoculum preparation across laboratories ensures reproducibility of results and meaningful comparisons across different studies and locations.

Researchers have also noted that carbon source availability, medium age, and bacterial growth phase influence virulence expression. Cells harvested during late exponential or early stationary phases tend to exhibit the most stable virulence phenotype. Additionally, inoculum water quality, pH balance, and ionic strength can impact bacterial survival and infection efficiency. Therefore, strict adherence to preparation protocols is critical in generating consistent and interpretable lesion data.

7.4 INOCULATION METHOD

Artificial inoculation is typically performed 30 days after transplanting or during the active tillering stage, a growth phase that provides optimal leaf development for lesion observation. The widely used leaf tip clipping method requires cutting approximately 3–5 cm of the leaf tip using sterilized scissors dipped into the bacterial suspension. This ensures immediate bacterial contact with exposed xylem vessels, facilitating rapid colonization. Each cut must be made using a single clean stroke to avoid tissue crushing that can alter lesion development.

To enhance pathogen entry, the clipped tips may be dipped directly into the inoculum suspension. This optional step improves infection uniformity, particularly when inoculum adherence is variable across leaf surfaces. After inoculation, maintaining adequate moisture around the leaves promotes bacterial entry and establishment. Covering inoculated leaves with polyethylene bags or shredded netting helps regulate temperature and humidity, preventing rapid drying of cut surfaces and ensuring consistent lesion progression.

Inoculation should be performed during cooler parts of the day, such as early morning or late evening, to minimize stress on the leaves from high midday temperatures. Heat stress can delay lesion onset, suppress bacterial multiplication, or lead to desiccation of the inoculated area, complicating disease assessment. Precautions must also be taken to avoid cross-contamination between plants, particularly in experiments involving multiple isolates with varying virulence profiles. Each set of scissors should ideally be used for a single isolate or sterilized thoroughly between samples.

7.5 OBSERVATION AND SCORING

Lesion assessment is conducted 21 days post-inoculation to allow sufficient time for progression and expression of differential host reactions. Lesions generally appear as straw-colored necrotic streaks extending from the cut tip downward along the leaf blade. Their progression may be linear or asymmetrical, occasionally following the midrib or one leaf margin more prominently than the other. Scoring involves measuring lesion length from the cut end to the furthest point of necrosis, regardless of direction or shape.

Accurate scoring requires careful identification of the three typical zones of lesion development: the necrotic, straw-colored region, the pale green water-soaked region just below it, and the healthy deep green tissue further down the leaf. Differentiating these zones is essential, especially in cases where lesion boundaries are irregular or diffuse. Standardized scoring scales categorize reactions as Resistant (<5 cm), Moderately Resistant (5–10 cm), Susceptible (10–15 cm), or Highly Susceptible (>15 cm). Ensuring uniform scoring across replications and observers is critical for reliable race classification. Figure 7.1 illustrates the typical lesion types and

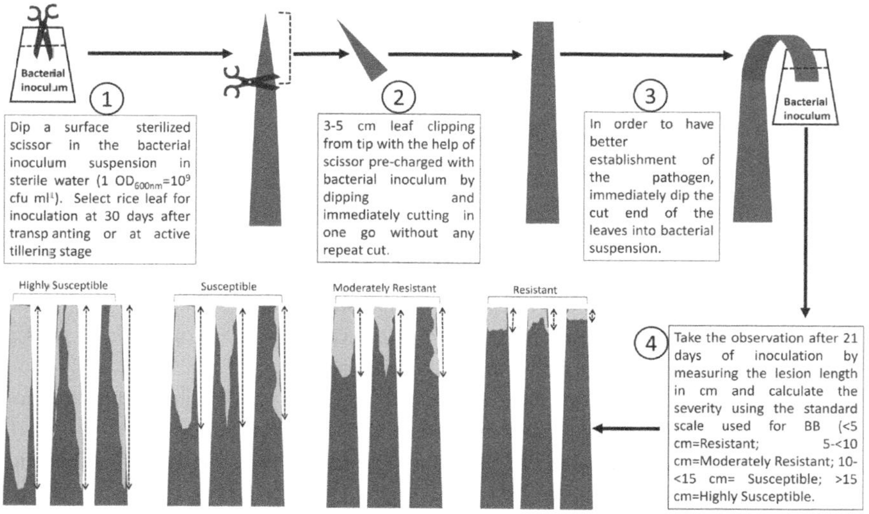

FIGURE 7.1 Leaf tip clipping technique for artificial inoculation of *Xoo*.

measurement methods used in bacterial blight studies. Consistency in measurement technique is essential for ensuring comparability of data across laboratories and seasons. Minor variations in measurement criteria can lead to major discrepancies in race assignment, especially in isolates showing intermediate virulence. Therefore, strict adherence to standardized protocols is essential for meaningful interpretation.

7.6 FACTORS AFFECTING ACCURACY OF ARTIFICIAL INOCULATION

Multiple factors can influence the accuracy and reproducibility of artificial inoculation results, often introducing variability that complicates race identification. Plant age is a key determinant, as tissues become physiologically more resistant with maturity, leading to underestimation of virulence if inoculation is performed too late. Similarly, environmental conditions such as humidity, temperature fluctuations, and sunlight exposure affect lesion initiation and expansion, potentially masking differences between isolates.

Inoculum density must remain uniform across replicates, as excessively high concentrations may cause lesion saturation that obscures differential reactions among NILs. Conversely, low densities may delay infection or result in aberrantly short lesions. The physiological state of the bacterium also plays an important role; cells harvested at inappropriate phases may express attenuated virulence factors, including reduced secretion of cell wall–degrading enzymes and TAL effectors. Variability in clipping depth, blade sharpness, and angle can result in inconsistent lesion initiation points, further influencing disease development.

Host plant nutrition and physiological stress have been shown to alter susceptibility responses, with nitrogen-rich plants often exhibiting enhanced lesion lengths. Field soil heterogeneity and microclimate variability can further impact disease expression. Additionally, slight contamination of inoculum or tools can produce mixed infections with altered virulence patterns. For these reasons, maintaining strict standardization of inoculation procedures, environmental conditions, and plant growth stages is essential for accurate race determination.

7.7 EVOLUTIONARY DYNAMICS OF RACE EMERGENCE

The emergence of new *Xoo* races is intimately linked to the evolutionary pressures exerted by widespread cultivation of rice varieties carrying specific resistance genes. Pathogen populations subjected to strong selection often evolve virulence traits that overcome these genes. Midha *et al.* (2017) noted that newly emerging races tend to exhibit increased aggressiveness and broader virulence spectra, reflecting their adaptive responses to host resistance deployment strategies. This ongoing arms race between host and pathogen is a hallmark of coevolution in agricultural ecosystems.

At the molecular level, the diversification of Transcription Activator-Like (TAL) effectors plays a central role in race evolution. TAL effectors manipulate host gene expression to promote susceptibility, and minor variations in their repeat structures can dramatically alter host range and virulence potential. Similarly, horizontal gene transfer, recombination, and small deletions or insertions in effector gene clusters contribute to rapid shifts in pathogenicity. Environmental factors and cropping system changes further accelerate selection for virulent strains by altering epidemiological patterns and transmission dynamics.

Race emergence also reflects long-distance dissemination patterns facilitated by seed movement, irrigation networks, and monsoon winds. Once introduced to a new region, a virulent race may expand rapidly if local cultivars lack effective resistance. This has been observed multiple times in South Asia, where introduction of new rice varieties has frequently been followed by virulence shifts in *Xoo* populations. Consequently, continuous monitoring of virulence trends is necessary to detect early signals of race emergence and to update resistance deployment strategies accordingly.

7.8 RACE SCENARIO IN MULTI-LOCATION SURVEYS

Multi-location surveys provide a more complete picture of *Xoo* race distribution across diverse agroecological zones. Data generated from multiple regions during different seasons offer insights into spatial patterns, temporal stability, and environmental determinants of race prevalence. Such surveys also capture the influence of host diversity, cropping patterns, and seed movement on pathogen population structure.

When isolates from multiple locations are screened on the same NIL set under controlled conditions, regional comparisons become possible, enabling identification of hotspots of virulence emergence. For instance, areas with intensive cultivation of specific *Xa* gene carriers may show rapid shifts toward races that overcome those genes. Historical datasets from IRRI and national programs indicate that race structures often shift dramatically within short periods, especially following the release of popular varieties carrying single major resistance genes.

Multi-location surveys also reveal the extent of race overlap among regions, which helps assess the potential for long-distance spread. In India, distinct clusters of races have been observed across northern, eastern, and southern rice-growing belts, reflecting both ecological and genetic factors influencing pathogen evolution. Integrating survey data with climatic and cropping information enhances the predictive power of race surveillance efforts, supporting informed breeding strategies and disease-risk assessments.

7.9 QUALITY CONTROL IN RACE IDENTIFICATION AND INTER-LABORATORY CALIBRATION

Accurate race identification is fundamentally dependent on rigorous quality control measures implemented consistently at every stage of the experimental workflow. Even minor sources of variability can lead to substantial differences in observed disease reactions, ultimately resulting in erroneous race classification. Human error during inoculation, scoring, or data recording, fluctuations in environmental conditions within growth facilities, and subtle inconsistencies in inoculum preparation are all potential contributors to misclassification. Because race identification forms the basis for breeding decisions, surveillance outputs, and epidemiological inferences, ensuring reproducibility and precision is of paramount importance.

Standardization of experimental protocols represents the first and most critical layer of quality control. Uniform plant growth conditions, including temperature, photoperiod, light intensity, humidity, and nutrient availability, must be carefully maintained to minimize physiological variation among host plants. Differences in seed age, seed source, or germination conditions can influence plant vigor and, consequently, disease expression. Similarly, inoculum preparation requires strict adherence to standardized procedures with respect to bacterial culture age, growth medium, inoculum density, and suspension homogeneity. Even slight deviations in bacterial concentration or inoculation volume can alter lesion development, particularly when evaluating near-isogenic lines that differ subtly in resistance responses.

Inoculation technique itself is a major determinant of reproducibility. Variations in clipping angle, pressure applied during inoculation, and the position of the cut relative to the leaf tip can influence pathogen entry and subsequent lesion progression. To mitigate such effects, laboratories must adopt clearly defined standard operating procedures and ensure that all personnel are trained to perform inoculations in a consistent manner. Routine refresher training for technical staff is essential, not only to reinforce correct techniques but also to harmonize interpretation of scoring criteria across individuals. Such training reduces inter-operator variability and promotes consistency in experimental execution and data interpretation.

Inter-laboratory calibration further strengthens the reliability of race identification by aligning methodologies and outcomes across different research facilities. Regional and national laboratories often operate under slightly different environmental conditions or infrastructure constraints, which can introduce systematic differences in disease expression. Calibration initiatives address these challenges by employing standardized reference isolates, common sets of host differentials, and shared experimental designs. By comparing results obtained from identical materials across laboratories, inconsistencies attributable to local conditions or procedural differences can be identified and corrected.

Regular inter-laboratory comparison exercises also foster communication and consensus among researchers. Differences observed in lesion length measurements, resistance classifications, or race assignments prompt critical examination of protocols and scoring practices. Factors such as growth chamber calibration, humidity regulation, light quality, seed source variability, and inoculation timing are often scrutinized during such exercises. Through collaborative refinement of methodologies, laboratories progressively harmonize their practices, resulting in more comparable and reliable race surveillance data at the national and regional levels.

Data transparency and meticulous record keeping constitute another essential component of quality control. Comprehensive documentation of isolate provenance, passage history, storage conditions, and revival protocols ensures traceability and reproducibility. Detailed records of inoculum preparation, including optical density measurements, dilution steps, and timing, allow for retrospective evaluation of experiments when discrepancies arise. Environmental parameters during plant growth and disease development, such as temperature ranges, relative humidity, and light conditions, should also be recorded systematically. Maintaining such detailed datasets enables researchers to identify sources of variation and to interpret results within their proper experimental context.

Over time, the accumulation of well-documented, high-quality data supports the development of robust pathogen monitoring systems. These systems provide reliable inputs for resistance breeding programs and disease management strategies, reducing uncertainty and enhancing confidence in race distribution analyses. Consistent quality control thus underpins the credibility and long-term utility of race profiling efforts.

7.10 INTERPRETATION OF LESION DEVELOPMENT AND PROGRESSION

Interpretation of lesion development and progression represents a critical and often challenging aspect of race identification. Lesions caused by *Xoo* typically originate at the site of inoculation and progress downward along the leaf blade, reflecting bacterial movement through the vascular tissues. Early stages of infection are characterized by water-soaked streaks that gradually transition into necrotic regions as host cells collapse. However, this canonical progression can be modified by a range of biological and environmental factors, necessitating careful and informed interpretation.

Distinguishing actively expanding lesions from older, static necrotic tissue is essential for accurate scoring. Actively expanding lesions often exhibit a water-soaked margin and a diffuse boundary with adjacent healthy tissue, whereas older lesions display sharp boundaries and dry, straw-colored necrosis. Failure to discriminate between these zones can lead to overestimation or underestimation of lesion length. Moreover, lesions do not always follow a strictly linear pattern. Side-arm lesions extending laterally from the main streak, lesions confined to the midrib, or patchy water-soaked zones may occur depending on host genotype and pathogen virulence.

Host physiological responses further influence lesion appearance and progression. Some near-isogenic lines may exhibit delayed lesion onset, reflecting effective early defense responses that temporarily restrict pathogen multiplication. In such cases, lesion expansion may accelerate at later stages once defenses are compromised. Other lines may show rapid initial lesion development followed by stagnation, indicating partial resistance mechanisms that slow pathogen spread without completely preventing infection. Recognizing these temporal patterns is crucial for correctly interpreting host–pathogen interactions and for assigning races based on functional outcomes rather than superficial lesion measurements.

Detailed observation and documentation of lesion morphology greatly enhance interpretative accuracy. Recording color transitions, boundary sharpness, tissue translucency, and progression rates over time provides a more nuanced understanding of disease development than single time-point measurements alone. Photographic records, when captured under standardized lighting and angles, further support consistent interpretation and enable retrospective comparisons among experiments and laboratories.

Environmental stress factors introduce additional complexity into lesion interpretation. High temperatures can accelerate host senescence or restrict bacterial movement, resulting in lesions that appear shorter or atypical. Low humidity may limit water-soaked symptom development, while excessive moisture can exaggerate lesion expansion. Nutritional status of the host also plays a significant role. High nitrogen fertilization, for example, can enhance tissue succulence and promote larger lesions even in genotypes that are otherwise resistant. Conversely, nutrient deficiency or drought stress may suppress lesion development, masking underlying susceptibility.

Given these influences, careful temporal monitoring of lesion development is essential. Repeated measurements at defined intervals allow researchers to distinguish genuine resistance responses from environmentally induced artifacts. Such longitudinal assessment improves confidence in race classification and reduces the risk of misinterpretation based on transient or misleading symptoms.

Integrating phenotypic and genotypic data provides a powerful framework for confirming race identity and resolving ambiguities inherent in lesion-based assays alone. Phenotypic evaluation captures the functional outcome of host–pathogen interactions under specific environmental conditions, reflecting the combined effects of pathogen virulence factors and host defense responses. Genotypic profiling, in contrast, reveals the molecular determinants of virulence, offering insights into the mechanisms underlying observed disease phenotypes.

Genotypic data, particularly profiles of transcription activator-like (TAL) effector genes and other virulence-associated loci, can clarify cases in which phenotypic responses are inconsistent or atypical. Isolates that produce similar lesion lengths on a set of host differentials may nonetheless differ in their effector repertoires, suggesting alternative strategies for host manipulation. Conversely, isolates with highly similar effector profiles may exhibit variable lesion development due to differences in environmental conditions, host physiological status, or experimental execution. By considering both datasets together, researchers can distinguish true biological variation from experimental noise.

The integration of phenotypic and genotypic information also enhances the resolution of race classification systems. Rather than relying solely on lesion length thresholds, race designation can be informed by underlying virulence gene content, improving consistency and biological relevance. This approach reduces the likelihood of misclassification and supports more accurate tracking of pathogen populations over time.

Beyond race confirmation, integrated datasets facilitate deeper insights into pathogen evolution and dissemination. Genomic markers enable the identification of clonal lineages and assessment of genetic relatedness among isolates collected from different regions. When coupled with phenotypic data, these analyses reveal whether the spread of disease is driven by the expansion of particularly virulent races or by independent emergence of similar virulence traits under comparable selection pressures.

7.11 INTEGRATING PHENOTYPIC AND GENOTYPIC DATA FOR RACE CONFIRMATION

Combining phenotypic and genotypic data enhances the accuracy of race confirmation by providing complementary perspectives on virulence. While lesion-based phenotyping defines the functional outcome of infection, genotypic profiling identifies the underlying determinants of virulence. Together, these approaches create a more robust framework for resolving ambiguous cases in which lesions do not fit classical patterns.

Genotypic data, such as TAL effector gene profiles, can clarify inconsistencies observed in phenotypic assays. For example, isolates showing identical lesion lengths may possess different effector gene sets, reflecting subtle differences in virulence strategies. Conversely, isolates with similar effector profiles may express variable lesion lengths due to environmental influences or plant physiological status. Integrating both datasets minimizes misclassification and supports more accurate race designation.

This integrated approach also facilitates tracking of pathogen evolution and spread across regions. Genomic markers help identify clonal expansion of virulent races, while phenotypic assays confirm their functional impact on host plants. Together, these insights inform breeding programs and enable more targeted deployment of resistance genes, enhancing long-term disease management.

7.12 APPLICATIONS OF RACE PROFILING IN RESISTANCE BREEDING AND DISEASE SURVEILLANCE

The continuous evolution of pathogen populations and the emergence of new virulent races necessitate a systematic understanding of race scenario and virulence profiles of the emergent races. Within this context, race profiling provides actionable insights that directly support resistance breeding, disease surveillance, epidemiological analysis, and policy-level decision-making. The applications of race profiling extend beyond academic research, playing a decisive role in safeguarding rice productivity and ensuring food security in regions prone to bacterial blight epidemics.

7.12.1 Role of Race Profiling in Resistance Breeding

Resistance breeding relies heavily on accurate knowledge of pathogen diversity and virulence patterns. Race profiling plays a central role in this process by enabling breeders to identify which resistance (*R*) genes remain effective against prevailing and emerging *Xoo* populations. Through systematic screening of near-isogenic lines (NILs), single-gene lines, and pyramided lines against a representative panel of *Xoo* isolates, breeders can classify pathogen races based on their compatibility or incompatibility with known *Xa* resistance genes.

Such screening efforts allow for precise determination of the durability and breadth of individual resistance genes. Certain *Xa* genes that were once highly effective may lose their efficacy over time due to pathogen adaptation, while others may continue to confer resistance across multiple environments and seasons. Race profiling thus provides empirical evidence on the longevity of resistance genes, helping breeders avoid reliance on genes that have already been overcome in farmers' fields.

In addition, race profiling supports the rational design of gene pyramids. By understanding the virulence spectrum of different races, breeders can select complementary resistance genes that target distinct pathogen mechanisms. This strategic stacking of genes enhances the likelihood of achieving durable and broad-spectrum resistance, as the pathogen must simultaneously overcome multiple defense barriers. Race profiling therefore bridges the gap between molecular resistance discovery and practical varietal development.

7.12.2 Guiding Deployment of Resistant Varieties

Beyond varietal development, race profiling informs decisions related to the deployment of resistant cultivars across diverse agroecological zones. The effectiveness of a resistance gene or gene combination can vary significantly depending on the local race composition of *Xoo* populations. Race profiling enables region-specific recommendations, ensuring that varieties released or promoted in a given area are well matched to the locally prevalent races.

This targeted deployment minimizes the risk of resistance breakdown and prolongs the useful life of resistance genes. It also prevents the unnecessary spread of varieties carrying ineffective genes, which could otherwise accelerate the selection of more aggressive pathogen races. In this manner, race profiling contributes to sustainable resistance management by aligning breeding outputs with field-level realities.

7.12.3 Applications in Pathogen Surveillance and Early Warning Systems

Pathogen surveillance programs increasingly rely on race profiling as a core component of disease monitoring and early warning systems. Regular collection and characterization of *Xoo* isolates from different rice-growing regions enable researchers and regulatory agencies to detect shifts in race composition and virulence patterns over time. Such temporal surveillance is critical for identifying the emergence of new or highly virulent races before they cause widespread epidemics.

Early identification of emerging races allows for timely interventions, including the acceleration of breeding programs targeting newly evolved virulence traits. Surveillance data can also prompt revisions in varietal recommendations, encouraging farmers to adopt cultivars that remain resistant under current pathogen pressures. In addition, race profiling supports the strategic implementation of integrated disease management practices, combining host resistance with cultural, biological, and chemical control measures.

By functioning as an early warning tool, race profiling reduces the lag between pathogen evolution and management response, thereby limiting yield losses and economic damage. Its integration into national and regional surveillance networks strengthens preparedness against bacterial blight outbreaks.

7.12.4 Contribution to Seed Certification and Quality Assurance

Race profiling plays an increasingly important role in strengthening seed certification and quality assurance programs, particularly in crops such as rice, where disease resistance is a critical determinant of productivity and stability. Certified seed systems are designed to supply farmers with planting materials that are genetically pure, agronomically superior, and resilient to major biotic stresses. Among these attributes, resistance to locally prevalent pathogen populations is especially important, as the use of susceptible varieties can rapidly undermine crop performance and exacerbate disease epidemics. By integrating race profiling into certification frameworks, seed regulatory authorities can ensure that the resistance claims associated with released and marketed varieties remain valid under field conditions.

The core objective of seed certification is to maintain varietal identity and performance across successive generations of seed multiplication and distribution. While genetic purity and physical quality have traditionally received primary emphasis, functional resistance to diseases such as bacterial blight has gained increasing attention. Race profiling provides an empirical basis for evaluating whether the resistance genes present in a variety are effective against the pathogen races currently circulating in a target production region. This shifts certification from a purely genetic or morphological exercise to a biologically informed process that reflects real-world disease pressures.

Incorporation of race profiling data into certification protocols allows authorities to assess resistance relevance rather than relying solely on historical performance data. Varieties that were resistant at the time of release may gradually lose effectiveness as pathogen populations evolve. Without periodic reassessment, such varieties may continue to be multiplied and distributed despite offering diminishing protection. Race profiling enables timely identification of this erosion of resistance, allowing certification agencies to update recommendations and prevent large-scale dissemination of seed that no longer meets resistance standards.

This application is particularly critical in regions where rapid shifts in race composition have been documented. Intensive rice cultivation, widespread adoption of a limited number of elite varieties, and favorable climatic conditions can accelerate pathogen evolution and lead to the emergence of new virulent races. In such contexts, static certification criteria are insufficient. Evaluating seed lots or parental lines against a representative panel of locally prevalent races provides a dynamic assessment of resistance effectiveness and ensures alignment between certified seed and current pathogen threats.

Seed lot evaluation using race profiling does not imply routine large-scale pathogenicity testing of every seed batch but rather strategic validation at key stages of seed multiplication. Breeder seed and foundation seed, which serve as the genetic base for subsequent seed classes, can be periodically screened against representative races. Confirmation of resistance at these early stages provides confidence that certified and truthfully labeled seed derived from them will retain functional resistance in farmers' fields.

Race profiling also supports the maintenance of resistance integrity during seed multiplication. Mechanical admixtures, genetic drift, or inadvertent selection can alter the resistance profile of a variety over successive generations. Periodic phenotypic screening against diagnostic races can reveal subtle shifts in host response that might otherwise go unnoticed. Early detection of such changes allows corrective action, including purification or replacement of seed stocks, thereby preserving varietal integrity.

From a quality assurance perspective, race profiling enhances transparency and accountability within seed systems. Documented evidence that a variety has been evaluated against relevant pathogen races strengthens the credibility of certification decisions and builds trust among farmers, extension agencies, and policymakers. Farmers are more likely to invest in certified seed when they have confidence that resistance claims are based on current and region-specific pathogen information rather than outdated trial data.

The integration of race profiling into seed certification also supports region-specific varietal recommendations. Different production zones may harbor distinct race compositions, resulting in differential performance of the same variety across regions. Certification authorities can use race profiling data to tailor recommendations, promoting varieties that are well matched to local pathogen populations while discouraging the use of less suitable ones. Such targeted deployment reduces disease risk and contributes to more stable production systems.

In addition, race profiling contributes to post-release monitoring of varietal performance. Feedback from surveillance programs can be linked with certified seed data to track how varieties perform as pathogen populations evolve. When resistance breakdown is detected, certification agencies can initiate timely reviews of varietal status, revise certification guidelines, or recommend phased withdrawal of affected varieties. This adaptive approach helps prevent sudden large-scale failures due to widespread cultivation of vulnerable cultivars.

Seed certification informed by race profiling also aligns closely with resistance breeding objectives. Data generated through certification-related screening can be shared with breeders, providing valuable insights into which resistance genes remain effective under field conditions. This feedback loop strengthens the overall breeding–deployment–certification continuum, ensuring that breeding outputs are rapidly translated into reliable planting materials for farmers.

At a broader level, incorporating race profiling into seed certification contributes to the sustainability of resistance management. By avoiding prolonged reliance on varieties with compromised resistance, certification systems help reduce selection pressure on pathogen populations and slow the emergence of highly virulent races. In this way, quality assurance based on race profiling not only protects individual farmers but also supports landscape-level disease management.

Ultimately, the contribution of race profiling to seed certification and quality assurance extends beyond technical validation of resistance. It reinforces farmer confidence in certified seed, promotes responsible varietal deployment, and enhances the resilience of rice production systems. As pathogen populations continue to evolve under changing agricultural practices and climatic conditions, the integration of race profiling into certification frameworks will become increasingly important for maintaining productivity, stability, and food security.

7.12.5 Insights into Epidemiology and Pathogen Evolution

In addition to its direct applications in resistance breeding and disease surveillance, race profiling contributes profoundly to epidemiological research and to the broader understanding of pathogen evolution. By systematically characterizing the virulence profiles of pathogen populations, race profiling provides a framework for analyzing how pathogens spread, diversify, and adapt in response to host resistance and environmental pressures. For bacterial blight of rice caused by *Xoo*, such insights are particularly important, given the pathogen's capacity for rapid evolution and long-distance dissemination.

Analysis of race distribution patterns across space and time allows researchers to infer key epidemiological processes underlying disease outbreaks. Spatial analysis of race composition reveals how different races are distributed across agroecological zones, irrigation networks, and rice production systems. Temporal monitoring, on the other hand, documents changes in race frequency and dominance over successive seasons or years, highlighting trends in pathogen adaptation. Differences in race composition among regions often reflect localized selection pressures. For instance, regions dominated by a narrow range of high-yielding varieties carrying similar resistance genes may exert strong directional selection on *Xoo* populations, leading to the rapid emergence of races capable of overcoming those genes. In contrast, areas with greater varietal diversity may harbor more heterogeneous pathogen populations but experience slower resistance breakdown. Race profiling thus serves as a window into the evolutionary consequences of cropping patterns and varietal deployment strategies.

Race profiling also aids in elucidating pathogen movement and population structure. The presence of similar races across geographically distant regions may indicate long-distance dissemination through seed exchange, irrigation water, or human-mediated transport. Conversely, the occurrence of unique or region-specific races can suggest localized evolution and limited gene flow between pathogen populations. When combined with molecular typing approaches, race profiling strengthens inferences about population structure and evolutionary history. Phenotypic race data complement genotypic information by linking genetic variation to functional outcomes in host–pathogen interactions. Together, these approaches enable researchers to distinguish between clonal expansion of successful races and the independent emergence of similar virulence patterns through convergent evolution.

Integrating race profiling data with geographical information systems (GIS), climatic variables, and agronomic data significantly enhances understanding of the ecological drivers that shape virulence evolution. Climate factors such as temperature, rainfall, humidity, and extreme weather events influence pathogen survival, reproduction, and dispersal, thereby affecting race dynamics. For example, warmer temperatures and high humidity may favor rapid disease development and increase opportunities for pathogen multiplication and mutation. Agronomic practices, including planting density, fertilizer use, irrigation methods, and crop residue management, further modulate disease pressure and selection intensity. Race profiling, when analyzed alongside these variables, can reveal how specific management practices contribute to the emergence or persistence of particular races. Such integrative analyses help identify environments that act as hotspots of pathogen diversity or as reservoirs for highly virulent races.

One of the most valuable contributions of race profiling to epidemiology is its ability to identify hotspots of pathogen diversity and corridors of race migration. Hotspots are regions where high race diversity is consistently observed, often due to intensive cultivation, continuous rice cropping, or the coexistence of multiple host genotypes. These areas are critical targets for intensified surveillance and proactive resistance management. Migration corridors, such as river basins, canal systems, or major seed distribution routes, facilitate the movement of races across landscapes. Mapping race occurrence along these corridors provides early indications of potential spread into new regions. Such information is crucial for anticipatory breeding and for implementing preventive measures before highly virulent races become established.

Insights gained from race-profiling–based epidemiological studies are invaluable for designing long-term disease management strategies. Understanding how races emerge, spread, and decline enables researchers and practitioners to predict future disease scenarios and to develop strategies that reduce selection pressure on resistance genes. These may include promoting varietal diversity, rotating resistance genes, and integrating host resistance with complementary management practices. Under changing climatic conditions, the importance of such foresight becomes even more pronounced. Climate change is expected to alter disease epidemiology by expanding the geographic range of pathogens and by modifying host–pathogen interactions. Continuous race profiling provides the empirical foundation needed to anticipate these shifts and to adapt management strategies accordingly.

7.12.6 Supporting Policy and Strategic Planning

At a broader scale, race profiling plays a pivotal role in informing policy decisions related to crop protection, agricultural sustainability, and food security. Policymakers require reliable, evidence-based assessments of pathogen threats to allocate resources effectively and to formulate strategies that balance productivity with resilience. Race profiling provides such evidence by translating complex pathogen dynamics into actionable risk assessments.

Evidence derived from race profiling enables policymakers to prioritize investments in resistance breeding programs, disease surveillance infrastructure, and human resource development. Regions identified as high-risk due to the presence of aggressive or rapidly evolving races may warrant increased funding for breeding initiatives and diagnostic capacity. Similarly, gaps in surveillance coverage revealed through race distribution analyses can guide the expansion of monitoring networks. Capacity building is another critical area supported by race profiling. Training programs for plant pathologists, breeders, and extension personnel can be tailored based on emerging pathogen threats and technological needs. Strengthening national expertise ensures that race profiling data are generated, interpreted, and applied effectively.

Race profiling data are central to national strategies for varietal replacement and deployment. As resistance genes lose effectiveness over time, continued cultivation of obsolete varieties can exacerbate disease pressure and accelerate pathogen evolution. Race profiling provides the scientific basis for identifying such varieties and for recommending their phased withdrawal. Conversely, newly developed varieties carrying effective resistance combinations can be strategically deployed in regions where they are most likely to perform well. This targeted approach maximizes the benefits of public and private investments in breeding while minimizing the risk of rapid resistance breakdown.

Incorporation of race profiling into seed systems and regulatory frameworks enhances the credibility and effectiveness of varietal release and certification processes. Regulatory authorities can use race profiling data to assess the relevance of resistance claims and to ensure that released varieties meet regional disease challenges. Such science-based regulation protects farmers from investing in varieties with compromised resistance.

The value of race profiling extends beyond national boundaries. *Xoo* populations do not adhere to political borders, and transboundary movement of races poses a significant challenge to disease management. International exchange of race profiling information facilitates regional cooperation and coordinated responses to emerging threats. Collaborative surveillance networks, shared databases, and harmonized race classification systems enhance the collective ability to detect and respond to new races. Such cooperation is particularly important in regions with shared river basins, similar agroclimatic conditions, or active seed trade. Race profiling thus serves as a unifying tool for collective action against bacterial blight.

Ultimately, race profiling contributes to broader food security strategies by strengthening the resilience of rice production systems. By enabling early detection of virulent races, guiding resistance deployment, and informing long-term planning, race profiling reduces vulnerability to disease-induced yield losses. Its integration with geographical, climatic, and agronomic data further enhances its relevance in the face of environmental change and intensifying agricultural demands.

Overall, race profiling functions as a critical interface between science and policy, linking resistance breeding, disease surveillance, epidemiological research, and strategic planning. By illuminating patterns of pathogen adaptation and movement, it enhances understanding of the evolutionary processes that threaten crop health. At the same time, it provides policymakers with the evidence needed to make informed decisions that protect farmers' livelihoods and national food supplies.

Ultimately, comprehensive and continuous race profiling is indispensable for safeguarding rice production in regions vulnerable to bacterial blight. As pathogen populations continue to evolve under changing agricultural practices and climatic conditions, the importance of race profiling will only increase, reinforcing its role as a foundational component of sustainable disease management and long-term food security.

8 *hrp* Gene Clusters and Their Role in *Xoo* Pathogenicity

The *hrp* gene cluster in *Xanthomonas oryzae* pv. *oryzae* (*Xoo*) represents one of the central determinants of the bacterium's ability to infect and colonize rice tissues. This cluster encodes the type III secretion system (T3SS), an indispensable virulence apparatus that translocates bacterial effector proteins directly into plant cells. Without the coordinated function of the *hrp/hrc* genes, *Xoo* is unable to initiate successful parasitism, establish infection foci in rice leaves, or suppress host immunity. Mutants lacking specific Hrp components fail to produce characteristic water-soaked lesions, indicating the absolute requirement of T3SS-mediated translocation events for pathogenicity (Wengelnik *et al.*, 1999; Rossier *et al.*, 2000).

The importance of the *hrp/hrc* cluster lies not only in its structural contribution to effector delivery but also in its highly integrated regulatory architecture. It interprets plant-associated environmental cues, orchestrates the assembly of the injectisome, regulates substrate specificity switching, and synchronizes the secretion of effectors with bacterial contact or proximity to host cells. The system is therefore both a mechanical and regulatory hub, establishing a direct interface between *Xoo* and rice cellular machinery.

Xoo has further refined its Hrp apparatus to overcome the unique physical and biochemical defenses of rice, such as the silicified cuticle, the phenolic-rich apoplast, and the tightly regulated immune surveillance pathways. Structural specialization within the HrpE pilus, the Hpa family chaperones, and several secretion-associated domains suggests continuous adaptation driven by host-specific selection pressures (Garita-Cambronero *et al.*, 2016; Furutani *et al.*, 2017). These adaptations highlight the cluster's evolutionary significance and centrality in the long-term success of *Xoo* as a specialized monocot pathogen.

8.1 OVERALL ARCHITECTURE OF THE *XOO* *HRP/HRC* GENE CLUSTER

8.1.1 Genomic Organization and Evolutionary Considerations

The *Xoo hrp/hrc* region is organized as a compact genomic island spanning approximately 30 kb, occupying a distinct chromosomal locus characterized by a GC content approximately 5–6% lower than the genomic average. This disparity strongly suggests an ancient horizontal acquisition event, followed by gradual assimilation and

DOI: 10.1201/9781003714590-8

specialization within the *Xanthomonas* lineage (Garita-Cambronero *et al.*, 2016). Over evolutionary time, the cluster has undergone extensive recombination, modular rearrangement, and gene accretion or loss events, resulting in a streamlined yet highly adaptable virulence island.

Comparative genomics shows that core *hrc* genes exhibit strong conservation across *Xanthomonas* species, reflecting the structural constraints of T3SS architecture. Essential components such as *hrcU, hrcV, hrcN, hrcQ, hrcJ,* and *hrcC* retain high synteny and sequence conservation (Jiang *et al.*, 2020). These genes encode the inner membrane export apparatus, cytoplasmic sorting platform, ATPase complex, and the outer membrane secretin, all of which are vital to the assembly and function of the T3SS injectisome.

In contrast, non-core genes—including *hrpE, hrpF, hrpD,* and several *hpa* accessory proteins—show signatures of positive selection. Elevated dN/dS ratios, lineage-specific insertions, and shortened or extended protein domains in these genes reflect the high evolutionary pressure imposed by the host plant. The variation in HrpE C-terminal antigenic domains suggests active diversification to evade recognition by rice pattern recognition receptors (Wang *et al.*, 2018). Likewise, the structural divergence observed in HrpF correlates with its function as a translocon protein interacting with rice plasma membranes.

Flanking regions of the *hrp/hrc* cluster harbor numerous small hypothetical ORFs. Although their functions remain unclear, their rapid evolutionary rate and proximity to key T3SS loci imply possible roles in regulatory fine-tuning, membrane remodeling during secretion, or modification of effector repertoires. The tight integration of these elements strengthens the concept of the *hrp/hrc* region as an evolving pathogenicity island under continuous host-mediated selection.

8.1.2 Operon Map and Expression Logic

The *hrp/hrc* cluster comprises several operons that are transcriptionally coordinated through PIP-box motifs—canonical cis-regulatory elements recognized by the transcription factor HrpX. The operons include:

1. *hrpG–hrpX* regulatory operon,
2. Multiple *hrc* operons encoding structural components,
3. *hrpE/hrpF* operon encoding the pilus and translocon,
4. *hpa* accessory operons,
5. Chaperone and effector-associated operons.

Each operon is tightly regulated to ensure proper timing of T3SS assembly, secretion hierarchy, and energetic efficiency. Under plant-inducing conditions, such as those mimicked by XOM2 medium, HrpG activates HrpX, which binds PIP-box elements upstream of the operons. HrpX binding results in a cascade of transcriptional activation culminating in the robust expression of pilus, translocon, and effector genes (Tsuge *et al.*, 2005).

An important architectural feature is the overlapping arrangement of promoters and terminators within the cluster. This overlapping creates a form of transcriptional "continuum," where strong HrpX induction leads to extended transcription across operonic boundaries. Consequently:

- Under low induction, only early structural genes are transcribed.
- Under full induction, entire blocks of *hrp*, *hrc*, and *hpa* genes are expressed simultaneously.

This flexible transcriptional strategy ensures that the bacterium can shift rapidly from basal T3SS assembly to high-efficiency effector secretion when host contact is established.

The operons share nearly identical transcriptional directionality, minimizing antisense interference. However, regulatory small RNAs encoded near operon junctions influence message degradation rates, contributing an additional post-transcriptional control layer that fine-tunes *hrp* gene expression in response to apoplastic nutrient availability, oxidative stress, or osmotic conditions.

8.2 *XOO*-SPECIFIC ADAPTATIONS WITHIN THE *HRP/HRC* CLUSTER

Although *Xanthomonas* species share a broadly conserved T3SS architecture, *Xoo* exhibits several distinct molecular and structural adaptations within the *hrp/hrc* gene cluster. These adaptations reflect long-term specialization to rice as a host and to the unique physical and physiological environments encountered during infection. Rice differs markedly from dicot hosts in surface composition, vascular structure, and immune signaling dynamics. As a result, *Xoo* has evolved modifications that fine-tune secretion efficiency, structural flexibility, and host interaction.

The *hrp/hrc* cluster of *Xoo* not only encodes the core secretion machinery but also incorporates lineage-specific variations that enhance survival and virulence within rice tissues. These adaptations are evident in pilus composition, periplasmic ring architecture, chaperone flexibility, and regulatory responsiveness. Together, they demonstrate how subtle structural changes can produce major functional advantages during host colonization.

8.2.1 Acidic Patch in the C-Terminal Region of HrpE

The major pilin subunit HrpE of *Xoo* contains a distinctive acidic patch at its C-terminal region. This patch is formed by clustered glutamate and aspartate residues and is notably absent in dicot-pathogenic *Xanthomonas* species. The presence of this acidic region represents a significant structural divergence that likely contributes to host-specific adaptation. One important consequence of the acidic patch is enhanced filament flexibility. Electrostatic repulsion between negatively charged residues can reduce the tight packing of pilin subunits. This results in a pilus that is less rigid and more capable of bending. Such flexibility is advantageous when navigating the narrow and uneven spaces of rice leaf tissues and the xylem environment.

Another predicted function of the acidic patch is reduced surface adhesion. Rice leaves possess a smooth, silicified cuticle that differs substantially from the rougher surfaces of many dicot hosts. Reduced nonspecific adhesion prevents the pilus from sticking prematurely to the leaf surface, allowing it to traverse the cuticle and reach target cells more effectively. The acidic patch may also modulate antigenicity and immunogenicity. Charged surface regions can influence how host immune receptors perceive bacterial structures. By altering surface charge distribution, *Xoo* may reduce recognition by rice immune surveillance systems, delaying activation of defense responses. Biophysical studies using NMR and AFM-based modeling support the idea that this C-terminal modification improves mechanical adaptability. These analyses suggest that the acidic region allows the pilus to withstand shear forces encountered in the apoplast and xylem. Such forces arise from water flow and pressure changes during transpiration.

In addition, the acidic patch may influence interactions with translocon components at the pilus tip. Subtle changes in electrostatic properties could improve alignment between the pilus and host membrane insertion machinery, thereby increasing the efficiency of effector delivery. Overall, the acidic patch in HrpE represents a fine-scale structural adaptation with broad functional consequences. It enhances flexibility, reduces unwanted adhesion, and may contribute to immune evasion, collectively optimizing pilus performance during rice infection.

8.2.2 Extended Structural Loop in HrcJ

HrcJ is a core periplasmic ring protein that plays a critical role in maintaining the structural integrity of the T3SS. In *Xoo*, HrcJ contains an extended loop domain that is not present in dicot-pathogenic relatives. This extension represents a lineage-specific modification with important functional implications. The periplasmic space of *Xoo* can expand under plant-like osmotic conditions. Such expansion alters the spatial relationship between the inner and outer membranes. The extended loop in HrcJ is predicted to accommodate this increased spacing, ensuring proper alignment of the secretion apparatus. Structural modeling suggests that the loop extension modifies the distance between the inner membrane ring and the periplasmic ring. This adjustment may stabilize the secretion apparatus under fluctuating osmotic conditions encountered during plant colonization. Without such flexibility, the injectisome could become misaligned or unstable. The extended loop may also act as a shock absorber. Mechanical stresses generated during pilus extrusion and effector translocation are transmitted through the periplasmic region. A flexible loop could dissipate these forces, preventing damage to the secretion system. Another possible function of the loop extension is improved interaction with neighboring T3SS components. By increasing contact surface area, HrcJ may form stronger or more adaptable interactions with other structural proteins, enhancing overall system robustness. This adaptation is particularly relevant in rice, where xylem sap composition and pressure vary dynamically. Maintaining a stable secretion apparatus under such conditions is essential for sustained virulence. The extended structural loop in HrcJ reflects adaptation to the unique physical environment of rice tissues. It enhances spatial flexibility, structural stability, and mechanical resilience of the T3SS in *Xoo*.

8.2.3 Expanded Coiled-Coil Regions in HpaB Family Chaperones

Xoo harbors distinctive HpaB family chaperones that contain longer coiled-coil regions than those observed in dicot-infecting *Xanthomonas* species (Furutani *et al.*, 2017). These structural expansions are believed to play a crucial role in managing the complex effector repertoire of *Xoo*. Coiled-coil domains are known to mediate protein–protein interactions. Expansion of these regions likely broadens the effector-binding spectrum of HpaB. This is particularly important for *Xoo*, which relies heavily on TAL effectors as well as multiple non-TAL effector families. The extended coiled-coil regions may allow HpaB to accommodate effectors with diverse sizes and structural properties. This flexibility ensures that a wide range of effectors can be maintained in a secretion-competent state. Another important function is the facilitation of hierarchical loading. The T3SS secretes substrates in a defined order. Expanded coiled-coil domains may help HpaB distinguish between early and late substrates, ensuring correct secretion timing. Specificity switching during secretion is also enhanced by chaperone flexibility. As the secretion system transitions from pilus and translocon assembly to effector delivery, chaperones must rapidly adjust binding preferences. Structural expansions likely support this dynamic switching. Given the central role of TAL effectors in *Xoo* virulence, efficient chaperone function is essential. Expanded HpaB flexibility may represent an evolutionary response to the need for precise and efficient TAL effector handling. Overall, expanded coiled-coil regions in HpaB chaperones illustrate how subtle structural changes can dramatically enhance secretion efficiency and adaptability, contributing directly to pathogenic success in rice.

8.2.4 Adaptations to the Silicified Rice Epidermis

The rice epidermis presents a uniquely challenging surface for bacterial pathogens. High silica content, smooth texture, and oxidative conditions require specialized adaptations for successful penetration and infection. *Xoo* exhibits multiple *hrp*/*hrc*-associated features that reflect these pressures.

High flexibility during pilus penetration is essential. Glycine-rich bends in HrpE allow the pilus to curve and adjust as it navigates the rigid rice cuticle. These bends reduce the risk of structural failure during penetration. Optimized surface adhesion is equally important. While excessive adhesion can hinder movement, controlled attachment is necessary for stable host contact. Modifications in pilus surface properties allow *Xoo* to balance these opposing needs.

The rice epidermis is also an oxidative environment. Enhanced resistance to oxidation is therefore critical. Structural reinforcement of translocon proteins, including strengthened HrpF domains, improves stability during membrane insertion under oxidative stress. Finely regulated secretion timing is another key adaptation. Premature effector release could trigger strong immune responses. *Xoo* exhibits refined HrpG/HrpX regulation that responds sensitively to apoplastic redox fluctuations, ensuring secretion occurs only under favorable conditions. Together, these features optimize T3SS performance during infection. They allow *Xoo* to penetrate the silicified epidermis, establish contact with host cells, and deliver effectors efficiently without triggering early defense responses.

Adaptations within the *hrp/hrc* cluster of *Xoo* reflect a sophisticated evolutionary response to the physical, chemical, and immunological challenges posed by rice. These modifications collectively enhance flexibility, stability, and regulatory precision, underpinning the success of *Xoo* as a vascular pathogen.

8.3 GENETIC ARCHITECTURE AND ORGANIZATIONAL LOGIC OF THE *HRP/HRC* GENE CLUSTERS

The *hrp/hrc* gene clusters of *X. oryzae* pv. *oryzae* (*Xoo*) constitute one of the most conserved and functionally integrated pathogenicity islands found among plant-pathogenic Gram negative bacteria. Their genetic organization reflects a finely tuned evolutionary design optimized for the secretion of effector proteins, translocation of virulence determinants into host cells, and survival within the nutrient-limited, defense-active environment of rice xylem. While the overall framework of the *hrp/hrc* cluster is broadly conserved across *Xanthomonas* spp., the specific arrangement, codon usage patterns, promoter structures, and regulatory elements reveal several lineage-specific refinements that underpin the pathogenic specialization of *Xoo* (Büttner and Bonas, 2010; Ryan *et al.*, 2011).

8.4 OVERALL COMPOSITION OF THE *HRP/HRC* CLUSTER IN *XOO*

The core *hrp/hrc* gene cluster in *Xoo* spans approximately 25–30 kb and encodes around 24–27 genes, depending on the strain and genomic background. These genes are organized into operonic modules responsible for (i) assembly of the T3SS basal body, (ii) formation of the extracellular pilus, (iii) effector secretion and translocation, and (iv) regulatory control of the secretion hierarchy (Furutani *et al.*, 2009; Noël *et al.*, 2001). Structurally, the cluster is typically organized into three contiguous regions: the HrpA operon, the HrpB operon, and the HrpC operon. Each of these regions encodes a specific set of components required to construct and operate the secretion system during host colonization.

8.4.1 THE *HRPA* OPERON: STRUCTURAL BACKBONE AND PILUS FORMATION

The *hrpA* operon encodes proteins central to pilus biogenesis and structural support. Among these, *hrpA1* encodes the major pilin subunit that polymerizes to form the extended extracellular pilus, a hallmark of T3SS architecture in *Xoo*. The pilus functions as a conduit for effector translocation and serves as the physical interface with rice host cells. Additional genes within this operon include *hrpB1* and *hrpB2*, which contribute to the assembly of the inner rod and basal ring, facilitating proper alignment between the bacterial cytoplasm and the host environment.

The molecular characteristics of *hrpA1* highlight its unique adaptation for long-distance effector delivery within the host xylem. Compared with related species such as *Xanthomonas campestris*, *Xoo hrpA* sequences display enhanced length conservation and strong purifying selection, suggesting functional constraints associated with successful colonization of rice vascular tissues (Cho *et al.*, 2008).

8.4.2 The *hrpB* Operon: Core Secretion Machinery and Conduit Assembly

The *hrpB* region constitutes the central secretion apparatus equivalent to the ESC or SctRSTU clusters described in animal-pathogenic T3SSs. Key components include HrcJ, HrcN, HrcQ, HrcR, HrcS, HrcT, and HrcU, all of which are conserved across plant-pathogenic bacteria (Büttner, 2012). These genes encode membrane-embedded ring-forming proteins, an ATPase complex responsible for energizing protein export, and scaffolding subunits that connect the cytoplasmic and periplasmic compartments.

Notably, *hrcN* encodes the T3SS ATPase that powers the unfolding and secretion of effector proteins. *Xoo hrcN* exhibits slightly modified regulatory motifs that promote rapid assembly during early xylem colonization, facilitating efficient deployment of effectors even at low bacterial density (Jha *et al.*, 2007). The *hrcQ–hrcU* module is essential for substrate specificity switching, a mechanism that ensures the correct secretion order: extracellular pilus subunits first, effectors later.

8.4.3 The *hrpC* Operon: Translocators and Secretion Regulators

The *hrpC* operon includes genes, such as *hrpF*, *hrpE*, *hrpG*, and *hrpX* (although HrpG and HrpX are often encoded adjacent but not strictly within the cluster). Among these, *hrpF* is particularly critical as it encodes a translocon protein that forms a membrane insertion pore complex in the host cell. The domains of *hrpF* in *Xoo* retain conserved helical regions essential for pore formation, yet also harbor strain-specific amino acid polymorphisms that may influence host recognition or specificity (Wengelnik *et al.*, 1999).

The *hrpE* gene encodes a small acidic protein involved in pilus tip assembly and is required for efficient effector translocation. Comparative studies indicate that *Xoo* HrpE contains unique acidic repeat signatures associated with flexibility and host-adaptive structural plasticity (Furutani *et al.*, 2009).

Though traditionally considered part of regulation rather than secretion, HrpG and HrpX form the master regulatory nodes controlling the cluster's expression. HrpG, a response regulator, activates HrpX, an AraC-type regulator that binds P_{hrp} promoters throughout the cluster. Their inclusion in this operonic region highlights the modular and hierarchical integration of regulatory control directly adjacent to structural gene modules (Guo *et al.*, 2012).

8.4.4 Promoter Architecture and Transcriptional Control Elements

Promoters within the *hrp/hrc* region exhibit several conserved features, including a canonical P_{hrp} box motif recognized by HrpX. This motif regulates expression in planta and under XVM2 or minimal medium that mimics xylem sap. The P_{hrp} box contains an imperfect palindromic sequence essential for functional binding, and mutations within these motifs drastically impair T3SS induction (Koebnik *et al.*, 2006).

Xoo isolates possess subtle yet functionally important promoter variations, enabling differential transcription intensity across strains adapted to diverse agro-ecological zones. These variations contribute to the observed heterogeneity in virulence outcomes and effector secretion efficiency.

8.5 CONSERVED AND VARIABLE REGIONS ACROSS XANTHOMONADS

While the basal machinery encoded by *hrc* genes is highly conserved across *Xanthomonas* species, the Hrp components, especially HrpA and HrpF, exhibit moderate sequence diversity. This distinction reflects the selective pressure exerted by host plants on extracellular and host-interfacing proteins, whereas intracellular structural proteins remain shielded from direct host recognition (Büttner and He, 2009).

Xoo-specific variations are most prominent in:

- effector-chaperone interaction motifs,
- translocon-encoding regions,
- promoter sequences linked to HrpX responsiveness, and
- pilus-associated proteins.

These adaptations highlight an evolutionary balance between maintaining essential secretion functions and enabling evasion of host surveillance mechanisms.

8.6 MOLECULAR MECHANISMS UNDERLYING T3SS ASSEMBLY AND SECRETION DYNAMICS IN *XOO*

The type III secretion system (T3SS) encoded by the *hrp/hrc* cluster in *Xoo* operates as a highly integrated nanomachine designed for the efficient deployment of effector proteins into rice cells. Beyond its structural organization, the system's functional success relies on a series of coordinated molecular events: hierarchical assembly of basal components, substrate recognition and secretion gating, pilus elongation, and the controlled activation of translocon insertion at the host interface. These processes unfold in a tightly regulated temporal order that ensures the accurate delivery of virulence factors precisely when and where they are required (Büttner, 2012; Guo *et al.*, 2012).

8.6.1 Hierarchical Assembly of the Basal Body and Inner Membrane Complex

Assembly begins deep within the bacterial cytoplasm, where core *hrc*-encoded proteins organize into a multi-ring basal structure embedded within the inner and outer membranes. The first elements to assemble are typically the inner membrane scaffolding proteins, such as HrcR, HrcS, and HrcT, which serve as anchors for subsequent recruitment of the export apparatus. These components form a stable foundation that supports the later insertion of membrane-spanning rings and the cytoplasmic ATPase complex (Noël *et al.*, 2001).

The basal body mirrors the architectural logic of flagellar T3SS systems, but in *Xoo* the components are optimized for effector delivery rather than motility. The membrane-spanning regions remain relatively rigid and act as conduits for the transfer of unfolded protein substrates. A notable feature in *Xoo* is the rapid assembly kinetics observed under *in planta*-mimicking conditions, suggesting that the bacterium invests heavily in early construction of the secretion platform upon sensing xylem fluid-like nutrients (Furutani *et al.*, 2006).

8.6.2 The ATPase Complex and Substrate Processing

A critical catalytic hub of the T3SS is the HrcN ATPase, positioned at the cytoplasmic base of the machinery. HrcN is responsible for energizing the unfolding of effector proteins and guiding them into the secretion channel. It recognizes secretion substrates through N-terminal secretion signals that lack a consensus sequence but contain loosely conserved hydrophobic and acidic amino acids.

Interaction between HrcN and chaperone–effector complexes dictates the initial tiers of secretion specificity. Chaperones such as Hpa family proteins bind effectors to maintain them in secretion-competent conformations and escort them to the ATPase. In *Xoo*, the affinity between effectors and their chaperones contributes to the finely tuned secretion hierarchy, in which early substrates (e.g., pilus subunits) are prioritized over late substrates (e.g., Xop effectors) (Furutani *et al.*, 2009).

8.6.3 Pilus Polymerization and Extracellular Extension

Once the basal body is established, the system transitions to pilus assembly. The pilus is principally composed of polymerized HrpA1 subunits. These subunits are directed through the secretion channel, where they undergo polymerization at the distal tip. The resulting filament extends outward from the bacterial surface and can reach substantial lengths, enabling the bacteria to bridge the physical distance to rice parenchyma or xylem pit membranes.

Xoo HrpA1 exhibits tailored biochemical features that enhance flexibility and resistance to shear forces expected in the xylem environment. Sequence variations among *Xoo* strains suggest adaptive evolution to maintain pilus integrity under fluid-flow conditions inside rice vascular tissues (Cho *et al.*, 2008). The pilus also functions as a structural scaffold that positions the translocon proteins at the host membrane, thereby acting as a delivery guide.

8.6.4 Translocon Assembly and Host Membrane Interaction

Following pilus extension, the system prepares for the deployment of translocon components such as HrpF and HrpE. Translocon assembly occurs only after the pilus provides a stable conduit, ensuring that pore formation is coordinated with effector secretion readiness. HrpF, in particular, inserts into the host's plasma membrane and forms a channel through which effectors ultimately enter host cells.

Structural analyses indicate that HrpF contains amphipathic helices that undergo conformational rearrangements upon contact with lipid bilayers. These helices are essential for membrane insertion and pore stabilization. In *Xoo*, small strain-specific polymorphisms in HrpF suggest that the protein may fine-tune its interaction with distinct rice genotypes, potentially contributing to host specificity patterns (Wengelnik *et al.*, 1999).

8.6.5 Secretion Hierarchy and Substrate Switching

The T3SS is not a passive conduit; it operates under a strict hierarchical programme known as secretion switching. Early substrates—such as HrpA1 pilus monomers—are secreted first. Only after the pilus is assembled does the machinery transition to secreting translocators, followed by effector proteins. This transition is mediated by components within the *hrcQ–hrcU* module and by regulatory interactions between the ATPase complex and chaperones (Büttner and He, 2009).

Effector proteins are exported in a largely unfolded state, but their secretion is timed carefully to avoid premature exposure to the extracellular environment. *Xoo* favours a secretion profile in planta that delivers effectors only after successful translocon formation, thereby minimizing wasteful or misdirected effector release.

8.6.6 Host-Triggered Activation and Environmental Cues

The T3SS of *Xoo* is highly responsive to plant-derived signals. Low-nutrient conditions, phenolic compounds, and components resembling xylem sap constituents activate the T3SS through the HrpG–HrpX regulatory cascade (Koebnik *et al.*, 2006; Guo *et al.*, 2012). Upon activation, HrpX binds to P_{hrp} promoters distributed across the *hrp/hrc* cluster, promoting transcription of structural genes, chaperones, and effector loci.

The system employs an efficient energy expenditure strategy: the T3SS is induced strongly only under conditions predictive of host association. Thus, *Xoo* avoids unnecessary metabolic burden while remaining prepared for rapid virulence deployment upon encountering rice tissues.

8.7 REGULATORY NETWORKS GOVERNING *HRP/HRC* GENE EXPRESSION IN *XOO*

The expression of the *hrp/hrc* gene cluster in *Xoo* is controlled by a multilayered regulatory framework that links environmental sensing, metabolic status, and host-derived cues to precise transcriptional responses. Unlike constitutively active virulence systems, the T3SS in *Xoo* is induced only under conditions predictive of host colonization, allowing the pathogen to balance metabolic cost with infection efficiency. This fine-tuned regulation is mediated through a combination of global regulators, two-component systems, AraC-family transcription factors, and nutrient-responsive modulators (Guo *et al.*, 2012; Koebnik *et al.*, 2006).

8.7.1 Central Role of the HrpG–HrpX Regulatory Cascade

The core regulatory axis of T3SS activation consists of the two key regulators HrpG and HrpX. HrpG, a response regulator of the OmpR family, functions as the primary sensor-integrator. Upon perception of host-associated signals such as phenolic compounds, osmoregulatory shifts, or low-nutrient conditions, HrpG transitions into an active conformation that drives the transcription of HrpX (Guo *et al.*, 2012).

HrpX, an AraC-type transcriptional activator, binds to the conserved P_{hrp} promoter motif located upstream of multiple *hrp/hrc* genes and effector loci. Activation of HrpX initiates a transcriptional cascade that simultaneously induces structural gene expression, effector synthesis, chaperone production, and secretion-associated proteins. This HrpG–HrpX axis thus forms the command center of T3SS expression.

A distinguishing feature of *Xoo* is the sensitivity of HrpX to subtle nutritional cues reflective of xylem sap. For example, minimal medium conditions that mimic low-carbon and low-nitrogen environments strongly upregulate *hrpX* expression, enabling rapid mobilization of the secretion system during early colonization (Ikawa and Tsuge, 2016).

8.7.2 P_{hrp} Promoters and Transcriptional Logic in the *hrp/hrc* Region

The P_{hrp} promoter motif recognized by HrpX is a defining regulatory element. It typically contains an imperfect palindromic sequence essential for HrpX binding. Each P_{hrp} box regulates a distinct subgroup of structural genes, translocon components, and effector-associated loci. The distribution of these motifs across the cluster creates a modular transcriptional network in which different operons respond with varying sensitivity to HrpX levels (Koebnik *et al.*, 2006).

This modularity allows *Xoo* to establish a temporal hierarchy of gene expression. Early-responding promoters induce assembly genes (e.g., *hrcQ*, *hrpB*, *hrpA*), whereas late-responding promoters regulate effector genes whose expression is required only after the secretion apparatus is fully assembled. Such staggered activation ensures that the bacterium avoids premature synthesis of energetically expensive effectors.

8.7.3 Integration of Nutritional and Metabolic Cues

Beyond the HrpG–HrpX axis, *Xoo* monitors its metabolic status to adjust T3SS expression. Limited availability of carbon sources such as sucrose or amino acids triggers pathways that converge on HrpG, amplifying its activation potential. Studies show that nutrient-poor conditions closer to xylem fluid composition significantly elevate transcription of *hrp* genes (Noël *et al.*, 2001).

In addition, evidence suggests the involvement of small regulatory RNAs that respond to carbon starvation or osmotic stress. These RNA-based regulators may modulate translation of HrpG, HrpX, or downstream transcripts, thereby fine-tuning secretion readiness under fluctuating xylem-like conditions. Although not fully characterized in *Xoo*, similar mechanisms in related *Xanthomonas* species support this model (Büttner and He, 2009).

8.7.4 Role of Two-Component Systems and Signal Transduction Networks

Several two-component systems (TCS) beyond HrpG contribute to the regulation of the T3SS. These TCS units sense variations in pH, osmolarity, redox status, or host-derived phenolic compounds and transmit these signals through phosphorylation cascades to regulatory nodes that interact with HrpG activity. For instance:

- osmosensory TCSs may enhance *hrp* expression when *Xoo* encounters the hypotonic environment of rice xylem,
- redox-sensitive systems can adjust secretion readiness when the pathogen experiences oxidative bursts in infected tissues.

These auxiliary TCS pathways ensure that T3SS activation is not driven by a single stimulus but by an integrated interpretation of multiple environmental cues relevant to infection progression.

8.7.5 Coordination between Virulence Regulators and Global Stress Pathways

The T3SS regulatory system functions in coordination with broader stress-response networks. Global regulators involved in quorum sensing, cyclic-di-GMP turnover, or stringent response interact indirectly with HrpG or transcriptional co-factors that modulate *hrpX* expression (Ryan *et al.*, 2011).

For example:

- low quorum-sensing signal levels in early infection may favor T3SS induction,
- high cyclic-di-GMP levels may repress motility while promoting secretion-related gene expression,
- ppGpp-mediated stringent response under nutrient limitation can amplify Hrp activation.

This crosstalk allows *Xoo* to synchronize T3SS activity with its physiological state, ensuring that virulence deployment aligns with metabolic feasibility and infection opportunity.

8.7.6 *Xoo*-Specific Refinements in Regulatory Architecture

Comparative genomics indicates that *Xoo* strains adapted to different rice-growing regions exhibit subtle regulatory variants—such as SNPs in P_{hrp} motifs, promoter length variations, and differences in HrpX responsiveness. These variations may contribute to differences in virulence intensity, effector secretion efficiency, and host genotype compatibility across *Xoo* populations (Furutani *et al.*, 2009).

Moreover, some *Xoo* strains carry additional regulatory inserts or intergenic expansions within the *hrp/hrc* region that may serve as binding sites for strain-specific transcription factors. These refinements illustrate ongoing evolutionary tuning of T3SS regulation in response to host diversity and environmental constraints.

8.8 STRUCTURAL AND FUNCTIONAL FRAMEWORK OF THE *HRC* CORE MACHINERY IN *XOO*

The *hrc* genes encode the conserved structural components that form the basal body, export apparatus, and secretion conduit of the type III secretion system (T3SS). In *Xoo*, these components constitute the architectural backbone that supports pilus formation, translocon deployment, and effector delivery. Although the overall organization resembles canonical T3SS systems, *Xoo* displays several host-adapted refinements that reflect its specialization for rice xylem colonization. The configuration of *hrc* genes ensures precise spatial alignment between inner membrane, periplasmic, and outer membrane modules, enabling the injectisome to operate efficiently despite the mechanical and biochemical constraints of the rice leaf environment.

8.8.1 Basal Body Organization and Spatial Coordination of Membrane Rings

The basal body is constructed from concentric ring complexes that traverse the bacterial envelope. In *Xoo*, the inner membrane ring composed primarily of *hrcD* and associated structural proteins provides the foundation for anchoring the export apparatus. This ring must align seamlessly with the periplasmic and outer membrane rings to form a continuous channel for substrate passage. Misalignment between rings, even by a few nanometers, disrupts secretion efficiency, indicating that inter-ring contacts are under strict evolutionary constraint.

The outer membrane ring is formed by the secretin protein encoded by *hrcC*. This multimeric secretin oligomerizes into a stable pore capable of withstanding the mechanical tension of pilus extrusion. Once assembled, the HrcC ring contributes to the rigid scaffold that supports the extension of the HrpE pilus through the bacterial envelope. Studies in related *Xanthomonas* species suggest that pilotin-like lipoproteins facilitate correct membrane insertion of HrcC, ensuring that the secretin pore is positioned at the correct angle relative to the inner membrane platform (Rossier *et al.*, 2000). *Xoo* likely employs a similar mechanism, although its specific pilotin counterpart has yet to be fully characterized.

Between these membrane rings lies a series of periplasmic cup-like assemblies. The HrcJ scaffold, supported by basal body components, bridges the gap between inner and outer rings. The spatial organization of HrcJ ensures that the pilus base and export channel remain correctly aligned during dynamic secretion events. Subtle variations in the periplasmic loop structures of *Xoo* HrcJ may enhance its flexibility, an adaptation that could support injectisome stability when the bacterium encounters fluctuating osmotic conditions in the rice xylem.

8.8.2 Composition and Operation of the Export Apparatus

Embedded within the basal body is the export apparatus, a complex of inner-membrane proteins responsible for mediating the passage of structural subunits, translocon proteins, and effectors. The export gate is primarily built from HrcV, HrcR, and HrcS homologs, working in concert with the substrate-switching protein HrcU. These protein complexes form a selective, partially gated channel that remains impermeable to folded proteins until ATPase-driven unfolding is initiated.

A defining feature of HrcU is its conserved NPTH autocleavage motif. Autocleavage of HrcU is required for the transition from early to late secretion stages—a functional switch that allows the export apparatus to prioritize pilus and translocon assembly before effectors are transported (Büttner *et al.*, 2006). *Xoo* appears to rely heavily on this cleavage-dependent switching, likely because its translocon deployment requires precise timing relative to the structural maturation of the HrpE pilus.

The export apparatus also includes regulatory gating elements that ensure substrates are secreted in a defined temporal order. Early substrates, typically Hrp and Hrc structural components, must traverse the gate before translocon proteins are admitted. Only after the translocon inserts into the host membrane is the apparatus committed to effector secretion. This programmed sequence prevents premature effector release into the apoplast, where they would be ineffective and energetically wasteful.

8.8.3 ATPase Complex and Mechanisms of Substrate Unfolding

At the cytoplasmic base of the injectisome resides the ATPase HrcN, a hexameric molecular motor structurally comparable to F1-ATPases. Its primary roles include substrate engagement, chaperone displacement, and partial unfolding of effector proteins prior to their entry into the secretion channel. Substrate recognition generally requires interaction between HrcN and chaperone–effector complexes, particularly those involving chaperones such as HpaB.

ATP hydrolysis induces a cyclical conformational change in HrcN that generates mechanical force. This force is transmitted to bound effector proteins, unfolding their N-terminal secretion signals and destabilizing secondary structures that would otherwise obstruct passage through the narrow export channel. In *Xoo*, this unfolding mechanism may be optimized for a large repertoire of structurally diverse Xop effectors, some of which contain unconventional chaperone-binding domains that interact with HrcN in a strain-specific manner (Furutani *et al.*, 2017).

The ATPase complex interacts with a cytoplasmic sorting platform that includes HrcQ-like components. These proteins create a dynamic scaffold that organizes incoming substrates into queues. By coordinating the order of substrate loading, the sorting platform ensures that the tightly regulated secretion hierarchy is maintained.

8.8.4 Hierarchy and Timing of Substrate Secretion

A hallmark of T3SS activity in *Xoo* is its strict secretion hierarchy. Structural proteins required for assembling the basal body and pilus are secreted first, followed by translocon components and finally effectors. The transition between these secretion stages is controlled by several molecular cues:

- early secretion is driven by high affinity between the export apparatus and core structural substrates,
- middle-stage secretion begins once pilus elongation reaches a threshold that signals readiness to deploy the translocon,
- late-stage secretion commences only after translocon insertion into the plant plasma membrane.

Proteins such as HpaB and HpaP regulate this transition by modulating substrate docking preferences. These regulators recognize specific N-terminal secretion signals and may change their binding affinity depending on the secretion stage. Some *Xoo* effectors appear to compete for shared chaperone binding sites, suggesting that effector abundance and chaperone occupancy contribute to secretion prioritization (Furutani *et al.*, 2017).

8.8.5 Quality Control and Assembly Surveillance

To ensure functional secretion, *Xoo* employs extensive quality control mechanisms centered on the *hrc* structural components. Misfolded pilin monomers, improperly assembled ring proteins, or stalled export intermediates can be targeted by cytoplasmic and periplasmic proteases. These include Lon and Clp-family proteases that degrade defective components before they accumulate and interfere with assembly.

In addition, some accessory proteins monitor the assembly status of the injectisome. For example, periplasmic surveillance proteins detect misaligned rings or incomplete secretin structures, triggering localized disassembly or repair. This ensures that the final machinery is both structurally stable and functionally competent—an essential requirement in the mechanically challenging rice xylem environment.

The interplay between chaperone systems, proteases, and assembly sensors highlights the energetic cost of T3SS deployment. As a result, *Xoo* invests heavily in ensuring that only fully functional injectisomes remain active during infection.

8.9 REGULATORY NETWORKS GOVERNING *HRP/HRC* GENE EXPRESSION IN *XOO*

The expression of the *hrp/hrc* gene clusters in *Xoo* is tightly controlled by a multilayered regulatory hierarchy that integrates environmental signals, host-derived cues, nutrient conditions, and intrinsic genetic modules. Because the T3SS is metabolically costly and essential only during host interaction, *Xoo* has evolved a finely tuned system to restrict *hrp/hrc* expression to plant-associated environments. This section provides a mechanistic overview of the major regulators, their interactions, and the broader signaling cascades that modulate T3SS deployment in *Xoo*.

8.9.1 The HrpG–HrpX Core Regulatory Cascade

8.9.1.1 HrpG: The Master Response Regulator

hrpG encodes an OmpR-family transcriptional response regulator that acts as the primary switch for initiating T3SS gene expression. Activation of HrpG is triggered under plant-mimicking conditions—low nutrient, acidic pH, and the presence of host cell wall-derived molecules. Once activated, HrpG undergoes a conformational transition that enhances its DNA-binding capacity, enabling it to directly induce HrpX, the key downstream transcriptional activator (Wengelnik *et al.*, 1999).

HrpG also mediates cross-talk with other virulence-associated systems, including extracellular polysaccharide synthesis and motility. This positions HrpG as the central integrative regulator that balances cellular strategies for host colonization.

8.9.1.2 *hrpX*: The AraC-Type Transcriptional Activator

hrpX encodes an AraC-type regulator that directly activates transcription of the structural *T3SS* genes, many *hrp* operons, and the majority of effector genes containing upstream *plant-inducible promoter (PIP) boxes* (Noël *et al.*, 2001). PIP box motifs (TTCGC–N_{15}–TTCGC) provide a hierarchical and coordinated mechanism for effector deployment.

Once HrpX is induced by HrpG, it triggers a cascade that includes:

- the *hrc* core apparatus genes,
- structural components of the secretion and pilus system,
- major chaperone and translocation-associated proteins,
- many Xop effector genes central to virulence.

Thus, the HrpG–HrpX cascade forms the backbone of the T3SS regulatory network.

8.10 ENVIRONMENTAL AND HOST-DERIVED SIGNAL INTEGRATION

8.10.1 Plant-Derived Metabolic Cues

Xoo relies heavily on chemical signatures released from rice tissues to gauge proximity to the host. These include:

- phenolic compounds,
- organic acids,
- cell wall hydrolysates,
- sugar derivatives produced during initial infection.

Such compounds relieve repression exerted by nutrient-rich media and allow HrpG/HrpX activation in a finely tuned manner (Büttner and Bonas, 2010).

8.10.2 Nutrient Limitation and Osmotic Conditions

Minimal-media environments resembling the apoplast induce T3SS expression. Limiting carbon and nitrogen sources mimics the restricted nutrient profile of the rice xylem, creating conditions that activate HrpG. Simultaneously, low osmotic potential and acidic pH, characteristic of xylem sap, further enhance the induction of T3SS genes.

8.10.3 Temperature and Oxidative Stress Cues

Although less studied in *Xoo*, cross-species analyses suggest that T3SS genes are sensitive to oxidative burst reactions and elevated temperatures. *Xoo* likely adjusts *hrp/hrc* activation in response to host-generated reactive oxygen species, enabling synchronized effector deployment during early infection stages.

8.11 ADDITIONAL TRANSCRIPTIONAL AND POST-TRANSCRIPTIONAL REGULATORS

8.11.1 Two-Component Systems beyond HrpG

While HrpG is the principal regulator, several additional two-component systems modify its activity. These auxiliary systems include:

- *ColR/S*, associated with membrane stress responses;
- *RpfC/G*, which responds to diffusible signal factor (DSF);
- *VemR*, a single-domain response regulator linked to virulence modulation.

Such regulators fine-tune HRP/HRC expression in response to dynamic *in planta* conditions (Buttner and He, 2009).

8.11.2 Small RNAs (sRNAs) and RNA-Binding Proteins

Increasing evidence suggests that sRNAs modulate *hrp/hrc* expression by controlling mRNA stability and translational efficiency. RNA-binding proteins like Hfq may stabilize regulatory sRNAs, influencing HrpX transcript abundance. This post-transcriptional layer helps calibrate T3SS gene expression in a rapid, reversible manner.

8.11.3 Global Metabolic Regulators (Clp, Fur, Zur)

Metabolic homeostasis intersects with virulence regulation. For instance:

- *Clp* modulates carbon metabolism and indirectly down-regulates T3SS when nutrients are abundant.
- Fur and Zur adjust virulence in response to iron and zinc availability, elements that fluctuate within host tissues.

Such regulators ensure that T3SS is activated primarily under conditions of metabolic stress typical of host environments.

8.12 QUORUM-SENSING AND DSF-MEDIATED REGULATION

8.12.1 Overview of DSF-Mediated Quorum Sensing in *Xoo*

Quorum sensing (QS) is a population density–dependent regulatory mechanism that enables bacterial communities to coordinate gene expression and collective behavior. In *Xoo*, the causal agent of bacterial blight of rice, QS is mediated primarily by diffusible signal factor (DSF) family fatty acid molecules. These signals regulate a wide spectrum of physiological and virulence-related traits, including motility, biofilm formation, extracellular enzyme production, and the type III secretion system (T3SS). Among these, the T3SS represents a key virulence determinant, as it facilitates the direct translocation of effector proteins into host cells to suppress immune responses and manipulate host cellular processes.

DSF-mediated QS in *Xoo* functions as an integrative signaling framework that links population density with environmental and host-associated cues. Rather than acting as a simple on–off switch, this system allows graded and context-dependent regulation of virulence, ensuring that T3SS expression is tightly controlled in accordance with infection stage and bacterial physiological status.

8.12.2 Molecular Link between DSF Signaling, c-di-GMP, and T3SS Regulation

The DSF signaling system in *Xoo* is encoded mainly by the *rpf* (regulation of pathogenicity factors) gene cluster. DSF molecules are synthesized by RpfF and accumulate extracellularly as bacterial cell density increases during host colonization. Signal perception occurs through the membrane-bound sensor kinase RpfC, which, upon DSF detection, initiates a phosphorylation cascade leading to activation of the response regulator RpfG. RpfG possesses an HD-GYP domain that functions as a phosphodiesterase, catalyzing the degradation of cyclic di-GMP (c-di-GMP).

Cyclic di-GMP acts as a central intracellular second messenger that integrates QS signals with global transcriptional regulation. In *Xoo*, elevated c-di-GMP levels are associated with repression of T3SS gene expression and promotion of sessile behaviors such as biofilm formation. Conversely, DSF-induced degradation of c-di-GMP relieves repression of the *hrp/hrc* regulatory cascade. Reduced c-di-GMP levels enhance the activity of the master regulators HrpG and HrpX, leading to HrpX-dependent activation of T3SS structural genes and effector-encoding loci. Through this mechanism, DSF-mediated QS directly connects extracellular population sensing with intracellular control of T3SS expression.

8.12.3 Population Density, Lifestyle Transitions, and Timing of T3SS Activation

A defining feature of quorum-sensing regulation in *Xoo* is the population density–dependent timing of T3SS activation. During early stages of infection, bacterial populations within the rice host are relatively low, resulting in limited DSF accumulation and elevated intracellular c-di-GMP levels. Under these conditions, T3SS expression is restrained, while traits associated with motility, surface attachment, and initial colonization are favored. This regulatory state minimizes premature exposure of immunogenic T3SS effectors to host defense systems.

As infection progresses and bacterial populations expand within the xylem, DSF concentrations increase and activate the RpfC/RpfG signaling pathway, leading to a decline in c-di-GMP levels. This shift marks a transition from a colonization-oriented lifestyle to an aggressive virulence phase characterized by enhanced T3SS assembly and effector secretion. High DSF levels have been correlated with increased effector delivery during early to mid stages of host colonization, when *Xoo* populations are sufficiently established to effectively suppress host immunity (Ryan *et al.*, 2015).

An important outcome of this regulation is the inverse relationship between biofilm formation and T3SS activation. High c-di-GMP levels promote extracellular polysaccharide production and stable biofilm development, which are advantageous for environmental persistence but incompatible with efficient T3SS function. DSF-mediated reduction of c-di-GMP shifts *Xoo* toward a more planktonic and invasive phenotype, facilitating T3SS-mediated host manipulation.

8.12.4 Biological Significance and Implications for Disease Management

DSF-mediated coordination of T3SS activity with population density and lifestyle transitions provides significant ecological and evolutionary advantages to *Xoo*. By restricting T3SS expression to conditions where bacterial populations have reached a critical threshold, the pathogen optimizes energy expenditure and reduces the likelihood of early immune recognition. This strategy enhances collective virulence while maintaining flexibility across diverse ecological niches, including epiphytic survival and systemic vascular colonization.

From an applied perspective, the quorum-sensing–T3SS regulatory nexus represents an attractive target for disease management strategies. Disruption of DSF signaling or interference with c-di-GMP metabolism offers a means to attenuate virulence without directly affecting bacterial viability, thereby reducing selective pressure for resistance development. Such anti-virulence approaches could complement host resistance breeding and contribute to sustainable management of bacterial blight of rice.

DSF-mediated quorum sensing plays a central role in regulating T3SS expression in *Xoo* by integrating population density, environmental conditions, and host-associated cues through the RpfC/RpfG–c-di-GMP signaling axis. This regulatory framework ensures precise temporal and physiological control of effector deployment, underscoring the importance of quorum sensing in the pathogenic strategy of *Xoo* (Ryan *et al.*, 2015).

8.13 EPIGENETIC AND CHROMATIN-LEVEL MODULATION

Bacteria do not possess classical chromatin structures. They do not have histones or nucleosome-based DNA organization. However, epigenetic regulation still plays an important role in *Xoo*. Among these mechanisms, DNA methylation is especially significant for regulating virulence-associated genes.

In *Xoo*, DNA methyltransferases modify specific nucleotide bases within the genome. These modifications do not alter the DNA sequence. Instead, they influence how easily regulatory proteins can access promoter regions. Changes in methylation can therefore affect gene expression patterns.

Promoter regions of *hrp/hrc* loci are particularly sensitive to methylation status. When these regions are methylated, binding of transcriptional regulators may be reduced. This can result in lower expression of genes involved in T3SS function. In contrast, reduced methylation may allow better access for regulatory proteins, leading to increased transcription.

Epigenetic regulation allows *Xoo* to respond rapidly to changing host conditions. Methylation patterns can change without permanent genetic alterations. This reversibility provides regulatory flexibility during infection, where environmental conditions fluctuate continuously. Phase-variable expression patterns have been observed in some *Xoo* isolates. In these cases, different cells within the same population show distinct expression states. Such variation may arise from differential DNA methylation. Some cells may express virulence genes strongly, while others remain in a low-expression state. This diversity improves population survival under stress.

Epigenetic modulation also contributes to long-term adaptation within the host. During systemic colonization, *Xoo* experiences sustained exposure to host defenses. Gradual changes in methylation patterns help fine-tune virulence output. This prevents excessive host immune activation and supports prolonged infection. Thus, although *Xoo* lacks classical chromatin, DNA methylation provides an important regulatory layer that complements transcriptional control and enhances adaptive flexibility.

8.14 CROSS-REGULATION WITH OTHER VIRULENCE PATHWAYS

T3SS activity in *Xoo* is closely connected with other virulence-related traits. These pathways do not function independently. Instead, they are coordinated through shared regulatory networks that ensure efficient infection.

Extracellular enzyme production is one such interconnected trait. *Xoo* secretes enzymes that degrade host macromolecules. These enzymes help release nutrients required for bacterial growth. The resulting nutrient-rich environment supports the energy demands associated with T3SS assembly and effector secretion.

Cell wall–degrading enzymes play an additional role beyond nutrient acquisition. Their activity releases host-derived molecules that may act as signals. These signals can stimulate the HrpG/HrpX regulatory cascade. In this way, damage to host tissues can directly reinforce virulence gene expression.

Motility and chemotaxis are also linked to T3SS regulation. During early stages of infection, motility helps *Xoo* reach entry sites such as hydathodes and wounds. At this stage, T3SS expression remains relatively low. Once attachment and colonization are established, motility decreases and T3SS activity increases. This coordinated transition supports efficient disease progression.

Biofilm formation shows an inverse relationship with T3SS deployment. Conditions that favor biofilm development generally repress T3SS activity. Conversely, when biofilm formation is reduced, T3SS expression increases. This balance allows *Xoo* to switch between survival-oriented and invasion-oriented lifestyles.

Through cross-regulation, *Xoo* ensures that virulence traits are activated in a logical sequence. Colonization, survival, and pathogenicity are tightly coordinated. This integration improves infection efficiency while conserving metabolic resources.

8.15 FUNCTIONAL IMPLICATIONS OF MULTI-LEVEL REGULATION

The presence of multiple regulatory layers provides *Xoo* with precise control over virulence expression. Regulation occurs at transcriptional, epigenetic, and signaling levels. Together, these mechanisms ensure accurate timing of T3SS activation. One important advantage is the prevention of premature virulence activation. Early expression of T3SS would require high energy input and could expose bacteria to host defenses. Multi-level regulation keeps virulence genes suppressed until conditions are favorable.

Effector deployment is spatially controlled. *Xoo* activates T3SS preferentially at infection sites such as hydathodes and wounds. These locations provide direct access to vascular tissues. Targeted activation improves infection success and limits unnecessary immune stimulation. Virulence regulation is also synchronized with host physiological changes. Variations in xylem sap flow and nutrient availability influence bacterial signaling pathways. *Xoo* responds by adjusting effector secretion accordingly. This synchronization supports systemic spread within the host.

Stress adaptation is another outcome of complex regulation. Host plants impose oxidative and osmotic stress during infection. Under such conditions, *Xoo* may reduce effector secretion to avoid detection. When stress levels decrease, virulence activity can resume.

Overall, multi-level regulation allows *Xoo* to balance aggression with restraint. This balance is essential for stable colonization of rice vasculature and long-term systemic infection.

8.16 STRUCTURAL ASSEMBLY AND FUNCTIONAL DYNAMICS OF THE TYPE III SECRETION SYSTEM IN *XOO*

The T3SS of *Xoo* is a complex multi-protein nanomachine. It enables direct delivery of effector proteins into rice cells. This delivery occurs without diffusion through the extracellular space, allowing precise manipulation of host processes. Although T3SS architecture is broadly conserved among Gram negative bacteria, *Xoo* exhibits lineage-specific features. These features influence secretion efficiency, pilus length, and interaction with host cells. Such adaptations reflect the vascular lifestyle of *Xoo*.

Assembly of the T3SS is tightly regulated. Components are produced in a defined order. This prevents wasteful assembly under non-infective conditions. Once assembled, the system forms a continuous channel connecting the bacterial interior to the host cell. Understanding the structure and dynamics of this system is critical for explaining how *Xoo* establishes infection and suppresses host immunity.

8.16.1 Overview of T3SS Architecture in *Xoo*

The T3SS resembles a molecular syringe. It is composed of three major structural modules. These modules work together as a single functional unit. The basal body anchors the system within the bacterial envelope. The secretion channel extends outward and forms a pilus-like structure. The cytoplasmic platform controls substrate recognition and secretion efficiency. The *hrc* genes encode the conserved structural components of the system. Several *hrp* genes contribute to pilus formation and translocation processes. Together, these components create a continuous conduit for effector transport. This architecture allows direct and efficient delivery of virulence proteins into host cells.

8.16.2 Basal Body: The Envelope-Spanning Core Structure

The basal body anchors the T3SS across the inner membrane, periplasm, and outer membrane. It forms a rigid cylindrical structure that stabilizes the entire apparatus. This structure shows similarity to the flagellar basal body, reflecting their evolutionary relationship (Galan and Collmer, 1999). The basal body provides mechanical support during secretion. It withstands forces generated by pilus extension and effector transport. At the same time, it maintains membrane integrity.

Assembly of the basal body follows an ordered sequence. Inner membrane components assemble first, followed by outer membrane elements. Proper assembly is essential for system functionality. Defects in basal body formation result in loss of secretion ability. This highlights its central role in virulence.

8.16.3 Inner Membrane Ring Components

The inner membrane ring is formed by proteins encoded by *hrcD, hrcR, hrcS*, and *hrcT*. These proteins assemble into a stable complex within the inner membrane. This complex creates a hydrophilic channel for substrate entry. Conserved residues line the channel and interact with secreted proteins. These interactions guide substrates toward the secretion pathway. The inner membrane ring is dynamic. It undergoes conformational changes that regulate channel opening and closing. This prevents leakage of cytoplasmic contents when secretion is inactive. Interactions with cytoplasmic regulators ensure that only properly recognized substrates are secreted.

8.16.4 Outer Membrane Secretin

The outer membrane channel of the T3SS is formed by HrcC. This protein belongs to the secretin family. It assembles into a large oligomeric pore in the outer membrane. In *Xoo*, HrcC forms a doughnut-shaped structure. This pore allows extrusion of the Hrp pilus into the apoplast. The structure supports pilus stability while maintaining membrane integrity. Proper assembly of HrcC is essential. Misassembly can disrupt the outer membrane and reduce virulence. Therefore, the outer membrane secretin plays a critical role in both structure and function of the secretion system.

8.17 THE HRP PILUS AND TRANSLOCON COMPLEX

The Hrp pilus and translocon complex together form the extracellular and host-interfacing part of the T3SS in *Xoo*. These structures enable the bacterium to deliver effector proteins directly from its cytoplasm into rice cells. This direct delivery is essential because plant cells are protected by a rigid cell wall and a plasma membrane, which together act as strong physical barriers. The pilus and translocon overcome these barriers by creating a continuous and protected passage for effector movement.

In the rice xylem, where *Xoo* primarily multiplies, the environment is highly dynamic due to constant water flow and pressure differences. The pilus–translocon system must therefore be both stable and flexible. The precise coordination between pilus assembly, translocon formation, and effector secretion allows *Xoo* to maintain efficient virulence even under these challenging conditions.

8.17.1 Pilus Structure and Polymerization

The extracellular pilus is mainly composed of HrpE, a small acidic protein that serves as the major structural subunit. HrpE monomers are secreted through the T3SS and assemble outside the bacterial cell into a long, hollow filament (Weber *et al.*, 2005). This filament can extend several hundred nanometers beyond the bacterial surface, allowing it to cross the plant cell wall and reach the host plasma membrane. The pilus has a helical structure that provides both strength and flexibility. Rigidity is necessary to prevent collapse under physical stress, while flexibility allows the pilus to bend and adjust its orientation within the complex plant tissue environment. This balance is particularly important in the rice xylem, where flowing sap can exert mechanical forces on bacterial cells and their surface structures. The inner diameter of the pilus is very narrow. Because of this, only unfolded or partially unfolded proteins can pass through it. Effector proteins of *Xoo* are therefore adapted to remain in a secretion-competent state during transit. This structural constraint ensures that secretion is tightly controlled and directional.

Pilus length is also regulated. Proper length ensures successful contact with the host membrane. Mutations affecting HrpE polymerization can lead to shorter or unstable pili, which severely reduce the efficiency of effector delivery and weaken virulence.

8.17.2 Translocon Formation at the Host Interface

At the distal end of the pilus, *Xoo* assembles a translocon complex at the host cell surface. This complex is primarily composed of HrpF and associated proteins. Once secreted, these proteins insert into the plant plasma membrane and form a gated pore. This pore acts as the final entry point for effectors into the host cytoplasm. Translocon formation occurs only after pilus assembly is complete. This timing is critical because premature membrane insertion could damage host cells without enabling effective effector transfer. The regulated deployment of the translocon ensures that membrane insertion and effector secretion are tightly coupled. The rice xylem presents unique physical conditions, including high water flow and negative pressure. These conditions could destabilize membrane-associated protein complexes. *Xoo* has adapted by strengthening interactions between translocon components and modifying their hydrophobic regions. These adaptations enhance membrane insertion and maintain pore stability during infection. The translocon is not a permanently open channel. Its gated nature allows selective passage of effector proteins while limiting uncontrolled ion movement across the host membrane. This reduces host cell damage and helps *Xoo* avoid early detection by plant immune systems.

8.17.3 Cytoplasmic Sorting Platform and ATPase Machinery

Inside the bacterial cytoplasm, the T3SS includes a sorting platform and an ATPase complex that coordinate substrate selection, unfolding, and secretion. These components ensure that secretion occurs in an ordered and energy-efficient manner.

8.17.3.1 Effector Recognition and Hierarchy

The sorting platform recognizes and organizes secretion substrates according to a defined hierarchy. Early substrates include proteins required for pilus assembly, followed by translocon components, and finally effector proteins. This order is essential for proper assembly and function of the secretion system. Effector proteins carry secretion signals, typically located at their N-termini. These signals are not strict sequence motifs but are defined by structural and biochemical features. Cytoplasmic chaperones, encoded by genes regulated by HrpG and HrpX, bind to effectors and guide them to the sorting platform. These chaperones also maintain effectors in a partially unfolded state, which is necessary for passage through the narrow pilus channel. In addition, chaperone–effector interactions help determine the order of secretion, allowing prioritization of effectors that suppress early host defenses.

8.17.3.2 ATPase-Driven Secretion

The cytoplasmic ATPase HrcN provides the energy required for secretion. HrcN hydrolyzes ATP and converts chemical energy into mechanical force. This force is used to unfold substrates and initiate their movement into the secretion channel. HrcN functions in a manner similar to F-type ATPases, using rotational movements to pull substrates toward the export gate (Akeda and Galán, 2005). ATP hydrolysis is tightly coupled with substrate recognition and gate opening. ATPase activity works together with the proton motive force across the inner membrane. This coordination ensures efficient and directional transport of proteins. Tight regulation of ATP consumption is important because excessive energy use would impose a metabolic burden on the bacterium.

8.17.3.3 Sequential Assembly of the T3SS Apparatus

Assembly of the T3SS occurs in a stepwise and highly regulated manner. This sequential process is essential to ensure that the secretion system is structurally stable, energetically efficient, and functionally precise. In *Xoo*, each assembly stage depends on the successful completion of the previous one. This strict order prevents wasteful protein secretion and avoids the formation of non-functional or harmful intermediates.

The assembly process begins inside the bacterial cell and progresses outward toward the host interface. Early steps focus on building a stable core within the bacterial envelope. Later steps involve the formation of extracellular structures and host-contact components. This inward-to-outward strategy allows *Xoo* to maintain envelope integrity while gradually extending the secretion machinery toward the host.

At the initial stage, inner membrane components assemble to form a stable foundation. These proteins create a platform that anchors the secretion system and defines the position where the T3SS will form. Without this anchor, subsequent components cannot be correctly aligned. This early stage is therefore critical for spatial organization of the entire apparatus.

Once the inner membrane structure is established, core export proteins are recruited. These proteins form the secretion channel that allows substrates to move from the cytoplasm toward the exterior. At this stage, the system remains inactive for effector secretion. This inactivity prevents premature energy consumption and avoids secretion before the apparatus is complete.

After stabilization of the inner membrane complex, assembly proceeds toward the outer membrane. The secretin component oligomerizes and inserts into the outer membrane, aligning precisely with the inner membrane channel. This alignment creates a continuous conduit across the bacterial envelope. Proper coordination between inner and outer membrane structures is essential, as misalignment would disrupt secretion and compromise membrane integrity.

During this intermediate phase, quality control mechanisms operate to ensure correct assembly. Improperly assembled components are not allowed to proceed to the next stage. This checkpoint-like regulation ensures that only fully functional structures are extended further. Such control is especially important in *Xoo*, where the T3SS must function under the physically demanding conditions of the rice xylem.

Once the envelope-spanning channel is completed, extracellular assembly begins. Pilus subunits are secreted through the newly formed channel and polymerize outside the bacterial surface. This extracellular assembly strategy avoids congestion within the cytoplasm and allows controlled elongation of the pilus. Pilus formation is gradual and tightly regulated to achieve optimal length. The pilus serves as a structural bridge between the bacterium and the host cell. Its assembly marks a transition from internal construction to host-oriented deployment. However, even at this stage, effector secretion does not begin. The system remains in a preparatory state until host contact is sensed.

Host contact acts as a critical signal for the final stages of assembly. Upon encountering rice tissue, *Xoo* initiates secretion of translocon proteins. These proteins travel through the completed pilus and insert into the host plasma membrane. This step completes the physical connection between the bacterial cytoplasm and the host cell interior. Only after translocon insertion does the T3SS become fully competent for effector delivery. This final activation step ensures that effectors are released only when a direct and functional pathway into the host cell exists. Such regulation minimizes loss of effectors into the extracellular environment and maximizes infection efficiency.

Throughout the assembly process, energy use is carefully controlled. ATP hydrolysis and proton motive force are engaged only when required. Early assembly steps require minimal energy compared to active secretion. This staged energy investment allows *Xoo* to conserve resources until infection conditions are favorable. The sequential nature of T3SS assembly also provides regulatory flexibility. Environmental signals, nutrient availability, and host-derived cues can pause or slow assembly at specific stages. This allows *Xoo* to adapt to changing conditions without committing fully to virulence.

Sequential assembly of the T3SS ensures that the secretion apparatus of *Xoo* is built accurately, activated at the correct time, and used efficiently. This ordered process integrates structural construction, energy management, and host sensing into a single coordinated pathway. Such precision is essential for successful colonization of rice tissues and sustained pathogenicity.

8.17.3.4 Early-Stage Inner Membrane Assembly

The early-stage inner membrane assembly represents the foundational step in construction of the T3SS in *Xoo*. At this stage, IMR proteins are inserted into the inner membrane and organize into a stable anchoring structure. This anchor defines the position and orientation of the secretion apparatus within the bacterial cell. Correct localization is essential because all later components must align precisely with this initial framework.

IMR proteins interact closely with the lipid bilayer of the inner membrane. These interactions stabilize the membrane and prevent leakage during later secretion events. The formation of this inner membrane platform ensures that the T3SS does not compromise cell viability while exporting large numbers of proteins. The IMR complex also serves as a recruitment hub. Core export components recognize and bind to this platform, allowing stepwise construction of the secretion channel. Without this recruitment function, the secretion machinery would remain incomplete and inactive.

Assembly at this stage is highly regulated. Only correctly folded IMR proteins are incorporated. Misfolded or improperly inserted components are excluded. This quality control mechanism prevents the formation of defective secretion systems that could waste cellular resources or damage membrane integrity. The timing of IMR assembly is also important. It occurs before the bacterium comes into close contact with host tissues. This allows *Xoo* to prepare its secretion machinery in advance, while still keeping it inactive. Such preparation ensures rapid activation once host-derived signals are detected. Energy requirements during early-stage assembly are relatively low. Most processes involve protein insertion and stabilization rather than active secretion. This allows *Xoo* to conserve energy while building the basic architecture of the T3SS. The IMR anchor also provides structural resilience. During later stages, mechanical stress generated by pilus extrusion and effector translocation is transmitted through this anchor. A weak or unstable IMR assembly would result in system failure. Therefore, this early stage plays a decisive role in determining overall functionality.

Early-stage inner membrane assembly establishes the physical and functional foundation of the T3SS in *Xoo*. It defines system location, ensures membrane stability, recruits essential components, and prepares the bacterium for subsequent stages of assembly and activation.

8.17.3.5 Outer Membrane Engagement and Secretin Polymerization

Following inner membrane assembly, the T3SS extends outward through engagement of the outer membrane. At this stage, the secretin protein HrcC oligomerizes to form a large pore within the outer membrane. This pore aligns precisely with the inner membrane channel, creating a continuous conduit across the bacterial envelope.

Secretin polymerization is a carefully controlled process. Individual HrcC subunits assemble into a ring-like structure that spans the outer membrane. Proper oligomerization is essential because incomplete or misaligned secretins would disrupt secretion and compromise membrane integrity. Alignment between the inner membrane complex and the HrcC pore is critical. This alignment ensures smooth passage of substrates without leakage into the periplasmic space. In *Xoo*, precise coordination between these components reflects tight regulatory control over assembly.

The outer membrane poses a unique challenge because it acts as a permeability barrier. The secretin must allow passage of large protein substrates while maintaining envelope stability. HrcC achieves this balance by forming a gated channel that remains closed until secretion is initiated. During this stage, the secretion apparatus is still inactive for effector export. The channel exists in a primed but closed state. This prevents premature protein loss and avoids triggering host defenses before contact with plant tissue. Environmental conditions influence secretin assembly. Factors such as membrane composition, stress signals, and nutrient status can modulate the efficiency of HrcC polymerization. This allows *Xoo* to adjust assembly according to surrounding conditions. The formation of a complete envelope-spanning channel marks a major transition in T3SS construction. At this point, the bacterium has established a secure pathway from the cytoplasm to the cell surface. Subsequent steps will extend this pathway toward the host. Overall, outer membrane engagement and secretin polymerization transform the T3SS from an internal structure into a trans-envelope system. This step is essential for both structural integrity and functional readiness.

8.17.3.6 Pilus Assembly and Extrusion

After formation of the membrane-spanning channel, pilus assembly begins. This stage represents the outward growth of the T3SS toward the host environment. HrpE monomers are secreted through the assembled channel and polymerize externally to form the pilus. Pilus assembly occurs outside the bacterial cell. This strategy prevents overcrowding within the cytoplasm and allows controlled elongation. Each HrpE subunit adds to the growing filament in an ordered manner. The pilus is hollow, forming a narrow channel through which effector proteins later pass. Its diameter restricts passage to unfolded or partially unfolded substrates. This structural constraint ensures selectivity and directional movement. Pilus length is tightly regulated. A pilus that is too short may fail to reach the host plasma membrane, while an excessively long pilus may become unstable. *Xoo* carefully balances these factors to optimize host contact. The mechanical properties of the pilus are crucial. It must withstand shear forces within the rice xylem while remaining flexible enough to navigate plant tissues. The helical structure of the pilus provides both strength and adaptability.

Mutations in HrpE disrupt pilus structure and impair T3SS function. Such mutations lead to reduced effector delivery and attenuated virulence, highlighting the essential role of pilus assembly in pathogenicity (Büttner *et al.*, 2002). Pilus extrusion also serves as a checkpoint. Only after a stable pilus is formed does the system proceed to deploy the translocon. This prevents incomplete systems from attempting effector secretion.

In essence, pilus assembly transforms the T3SS into a host-reaching structure. It bridges the physical gap between bacterium and plant cell, preparing the system for direct interaction with host membranes.

8.17.3.7 Translocon Deployment

The final assembly step of the T3SS is translocon deployment. This step occurs only after successful pilus assembly and host contact. Upon encountering rice tissues, *Xoo* secretes translocon proteins that travel through the pilus and insert into the plant plasma membrane.

The translocon forms a gated pore in the host membrane. This pore creates a direct connection between the bacterial cytoplasm and the host cell interior. Through this connection, effector proteins are delivered with high precision.

Translocon deployment is tightly regulated. Proteins are secreted only when host contact is confirmed. This prevents unnecessary membrane damage and reduces the risk of activating strong host immune responses prematurely. The rice xylem presents physical challenges such as high water flow and pressure. The translocon must therefore be stable and securely anchored in the host membrane. *Xoo* achieves this by modulating the hydrophobic properties of translocon proteins. Once inserted, the translocon does not remain permanently open. Its gated nature allows controlled passage of effectors while limiting ion leakage. This protects host cell integrity and prolongs bacterial survival.

Completion of translocon deployment marks full activation of the T3SS. At this stage, the system functions as an integrated unit composed of the inner membrane anchor, outer membrane channel, pilus, and host membrane pore. With the secretion pathway fully assembled, *Xoo* can efficiently deliver effectors that suppress host defenses, manipulate host metabolism, and promote systemic infection. This coordinated assembly ensures maximum virulence with minimal energy waste. Translocon deployment represents the final and decisive step in T3SS assembly. It converts a structural apparatus into an active virulence machine, enabling *Xoo* to establish successful infection within rice tissues.

8.18 MECHANISMS OF EFFECTOR TRANSLOCATION INTO PLANT CELLS

Effector translocation is the central functional outcome of the T3SS in *Xoo*. This process involves precise recognition of effector proteins, their controlled movement through the secretion channel, and their final delivery across the host plasma membrane. Each step is tightly regulated to ensure that effectors reach their intracellular targets efficiently and at the correct stage of infection.

8.18.1 Recognition of Secretion Signals

Effector proteins of *Xoo* carry secretion signals located at their N-termini. Unlike classical signal peptides, these signals do not share a conserved amino acid sequence. Instead, they are recognized through structural and physicochemical features such as amino acid composition, flexibility, and lack of stable secondary structure. Because the secretion signals are degenerate, specificity is achieved through interaction with cytoplasmic chaperones. These chaperones bind effectors shortly after synthesis and guide them to the T3SS sorting platform. The formation of chaperone–effector complexes ensures accurate substrate recognition and prevents incorrect proteins from entering the secretion pathway.

Recognition is also hierarchical. Some effectors are prioritized for early secretion, while others are exported later. This hierarchy allows *Xoo* to deploy effectors in a strategic sequence, first suppressing host immunity and then modifying host physiology to support bacterial multiplication. In addition, chaperones protect effectors from degradation and premature folding. This protection is critical because folded proteins cannot pass through the narrow secretion channel. Thus, secretion signal recognition is closely linked to effector stability and secretion competence.

8.18.2 Passage through the Channel

After recognition, effectors are directed to the secretion channel and begin their passage through the T3SS. The channel includes the inner membrane export gate, the envelope-spanning core, and the extracellular pilus. Throughout this journey, effectors remain in a transiently unfolded or partially unfolded state. The narrow lumen of the pilus imposes strict size constraints. Fully folded proteins cannot pass through it. To overcome this limitation, the T3SS ATPase and associated chaperones actively maintain effectors in a flexible conformation. ATP hydrolysis provides the energy required to unfold substrates and initiate their movement.

Passage through the channel is rapid but highly controlled. Effectors move in a linear fashion, minimizing exposure to the periplasmic space or extracellular environment. This protected transit reduces the risk of effector degradation and prevents unintended activation of host defenses. The coordination between ATPase activity, proton motive force, and channel gating ensures directional transport. This coordination allows *Xoo* to deliver large numbers of effectors efficiently during infection.

8.18.3 Host Membrane Penetration

The final barrier to effector delivery is the host plasma membrane. Penetration of this membrane is achieved through the translocon, which forms a selective pore at the host–pathogen interface. The translocon opens only in response to direct physical contact with plant membranes.

Host membrane properties strongly influence translocon insertion. Lipid composition, membrane fluidity, and membrane potential all affect the efficiency of pore formation. This sensitivity highlights the close co-evolution between *Xoo* and rice. The translocon functions as a gated channel. It allows passage of effector proteins while restricting uncontrolled ion flow. This selectivity preserves host cell viability long enough for effectors to manipulate host processes. Successful membrane penetration completes the translocation process, delivering effectors directly into the host cytoplasm where they exert their functions.

8.19 FUNCTIONAL IMPLICATIONS OF T3SS ARCHITECTURE IN *XOO* PATHOGENESIS

The architecture of the T3SS is not merely structural but directly shapes the pathogenic strategy of *Xoo*. Its design reflects adaptation to the rice vascular system, host immune pressure, and the need for precise temporal control of effector delivery.

8.19.1 Adaptation to Xylem Colonization

The rice xylem is a challenging environment characterized by continuous water flow and negative pressure. The extended pilus and robust translocon of *Xoo* allow effector delivery under these conditions. The length of the pilus enables contact with host cells despite physical separation caused by xylem sap flow. The mechanical stability of the translocon ensures sustained effector injection during systemic spread. This adaptation is essential for long-distance movement within the vascular system and for establishment of systemic disease.

8.19.2 Efficient Immune Suppression

Direct injection of effectors allows *Xoo* to rapidly suppress plant immune responses. By bypassing extracellular recognition, effectors can interfere with PTI and ETI signaling pathways at an early stage. Rapid immune suppression is particularly important in resistant rice varieties, where defense activation is swift. The efficiency of T3SS-mediated delivery allows *Xoo* to overcome these barriers and establish infection.

8.19.3 Co-Evolution with Host Surveillance

The structural plasticity of the T3SS reflects long-term co-evolution with rice immune systems. Changes in pilus composition and translocon proteins likely represent adaptations to avoid host recognition. Selective pressures imposed by host surveillance systems drive diversification of secretion components, allowing *Xoo* to maintain pathogenicity despite evolving host defenses.

8.19.4 Contribution of *hrp/hrc* Genes to Effector Translocation and *Xoo* Pathogenicity

Effector delivery in *Xoo* depends entirely on the coordinated action of *hrp* and *hrc* genes. These genes encode the structural framework, regulatory elements, and substrate specificity determinants of the T3SS. Because disease development requires a functional T3SS, the *hrp/hrc* cluster constitutes the central pathogenicity island of *Xoo*. Disruption of any major module within this cluster leads to loss of effector translocation and loss of pathogenicity.

8.19.5 Coordination between HrpG–HrpX Regulation and Effector Translocation

The coordination between regulatory control and physical secretion is central to the success of the T3SS in *Xoo*. The HrpG–HrpX regulatory cascade ensures that effector translocation occurs only when the secretion apparatus is fully assembled and environmental conditions are appropriate. This coordination prevents wasteful expression of virulence proteins and ensures precise timing of effector delivery during infection.

8.19.5.1 Activation of Secretion Competence

Upon perception of plant-derived signals, the response regulator HrpG becomes activated and induces expression of HrpX. HrpX functions as a transcriptional activator of PIP-box–containing genes. These genes encode the structural and functional components of the T3SS. This transcriptional program follows a strict temporal order. First, genes encoding basal body and inner membrane components are induced. These components establish the physical framework of the secretion system. Next, genes encoding pilus and translocon proteins are expressed, allowing extension of the apparatus toward the host. Only after these steps are completed are effector-associated genes activated (Wengelnik *et al.*, 1999; Tsuge *et al.*, 2005).

This sequential induction ensures that effectors are produced only when a functional secretion pathway exists. Without such regulation, effectors would accumulate in the cytoplasm without being secreted, leading to metabolic burden and possible toxicity. Thus, the HrpG–HrpX system does not merely control gene expression. It imposes the correct order of assembly and activation required for successful effector translocation.

8.19.5.2 Linking Gene Induction to Secretion Status

In *Xoo*, regulatory control is closely linked to the physical status of the T3SS. Evidence suggests that the secretion apparatus feeds back into the regulatory network. When basal body assembly is incomplete or pilus formation is disrupted, induction of PIP-box–containing genes declines.

This feedback mechanism acts as a surveillance system. It ensures that effector genes are not overexpressed when secretion cannot occur. Such regulation conserves energy and prevents the accumulation of unused virulence proteins.

By linking transcriptional induction to secretion competence, *Xoo* achieves tight coordination between regulation and function. This integration enhances pathogenic efficiency and adaptability during infection.

8.19.6 Roles of *hrc* Structural Genes in Effector Export

The *hrc* genes encode the conserved structural core of the T3SS and are indispensable for effector export in *Xoo*. These genes form the backbone of the secretion apparatus and define the pathway through which effectors move from the cytoplasm to the host cell.

The proteins encoded by *hrc* genes assemble into a continuous conduit that spans the inner membrane, periplasm, and outer membrane. This conduit is essential for maintaining directional and protected movement of effector proteins.

Each *hrc* gene product performs a specific structural or functional role. Together, they ensure proper assembly, stability, and activity of the secretion system. Loss of any major *hrc* component disrupts effector translocation and abolishes pathogenicity.

Because these genes are conserved across many Gram negative pathogens, they represent the fundamental machinery of the T3SS. In *Xoo*, their precise coordination reflects adaptation to plant-associated lifestyles and xylem colonization.

8.19.7 Inner Membrane Export Gate (hrcU, hrcV, hrcS)

The inner membrane export gate is formed primarily by the products of *hrcU, hrcV,* and *hrcS*. These proteins create the entry point for substrates into the T3SS channel.

HrcU plays a critical regulatory role. It undergoes NPTH autocleavage, which switches substrate specificity from early substrates, such as pilus and translocon components, to late substrates, namely effector proteins (Büttner *et al.*, 2006). This cleavage event acts as a molecular switch that determines when effector secretion can begin.

HrcV and HrcS contribute to the physical structure of the export gate. They help form a channel through which unfolded effectors enter the secretion pathway. These components also interact with cytoplasmic sorting proteins to ensure correct substrate selection. If substrate switching fails, effector secretion cannot occur even when the pilus and translocon are fully assembled. This highlights the central role of the export gate in coordinating structure with function.

8.19.8 Sorting Platform (HrcQ, HrcR, HrcN)

The sorting platform is located on the cytoplasmic side of the T3SS and serves as the control center for effector export. It integrates substrate recognition, unfolding, and energy transduction into a single coordinated process.

HrcQ binds chaperone–effector complexes and positions them at the export gate. HrcR assists in substrate discrimination, ensuring that only appropriate proteins are secreted at a given stage.

HrcN functions as the ATPase of the system. It hydrolyzes ATP to provide the energy required for effector unfolding and entry into the secretion channel. This ATP-driven process ensures that effectors remain flexible enough to pass through the narrow pilus.

Mutations in *hrcN* completely block effector secretion and eliminate pathogenicity. This demonstrates that the sorting platform is essential not only for efficiency but also for the basic functionality of the T3SS in *Xoo*.

8.19.9 Periplasmic and Outer Membrane Components (HrcJ, HrcC)

HrcJ and HrcC form the periplasmic and outer membrane elements of the T3SS. These components ensure continuity of the secretion pathway across the bacterial envelope.

HrcJ aligns the inner and outer membrane rings and stabilizes the periplasmic region of the apparatus. Proper alignment is essential for preventing leakage and maintaining directional transport. HrcC forms the outer membrane secretin pore. Through this pore, the pilus is extruded, and substrates pass toward the host. HrcC must balance permeability with membrane integrity. Any collapse or misalignment of these components renders the secretion pathway nonfunctional. Thus, HrcJ and HrcC are critical for maintaining structural coherence of the T3SS.

8.19.10 Contribution of *hrp* Structural Genes to Effector Translocation

Beyond the conserved *hrc* core, the *hrp* genes encode specialized structures that mediate the final stages of effector delivery. These include the pilus and translocon components. The *hrp*-encoded pilus enables physical contact with host cells, while translocon proteins create the pore through which effectors enter the host cytoplasm. These structures are essential for bridging the gap between bacterium and plant cell. The *hrp* genes also contribute to host specificity. Variations in pilus and translocon composition may reflect adaptation to rice tissues and immune pressures. Together, *hrp* and *hrc* genes form a fully integrated secretion system. Their coordinated action allows *Xoo* to deliver effectors precisely, efficiently, and in synchrony with host interaction, ultimately determining pathogenic success (Table 8.1).

8.19.11 The HrpE Pilus as the Conduit for Effector Passage

The *hrpE* gene encodes the major pilin subunit that polymerizes to form the extracellular pilus of the T3SS. This filamentous structure represents the only continuous physical connection between the bacterial cytoplasm and the host cell surface, making it indispensable for effector translocation in *Xoo*. Unlike passive secretion systems, the T3SS pilus functions as an active, mechanically responsive conduit that integrates structural, regulatory, and signaling roles.

A primary function of the HrpE pilus is to provide a direct physical route for effectors to exit the bacterial envelope. Once effectors pass through the inner membrane export gate and the periplasmic channel, they enter the narrow lumen of the pilus. The internal diameter of the pilus is highly constrained, allowing passage only of

TABLE 8.1
Role of *hrp/hrc* Genes in Effector Translocation

Role	*hrp/hrc* Genes	Specific Contribution
Regulatory induction	*hrpG, hrpX*	Ensures correct timing of secretion
Export initiation	*hrcU, hrcV, hrcS*	Channel opening and substrate switching
Sorting and unfolding	*hrcQ, hrcR, hrcN*	Effector recognition and unfolding
Pilus assembly	*hrpE*	Conduit for effector passage
Membrane penetration	*hrpF*	Translocon pore formation
Secretion refinement	*hpa1, hpaB, hrpD6*	Prioritization and stabilization

proteins that are maintained in an extended or partially unfolded conformation. This architectural constraint ensures directionality and specificity of secretion. Beyond acting as a passive conduit, the pilus serves as a dynamic scaffold for pilus extension. Continuous polymerization of HrpE monomers at the base enables elongation of the filament until host contact is achieved. This dynamic growth is tightly coordinated with secretion activity, ensuring that pilus assembly and effector translocation proceed in a synchronized manner. The pilus also plays a role in transmitting mechanical signals of host contact back to the bacterium. Physical resistance encountered upon contact with the rice cell surface generates mechanical feedback that is thought to influence secretion switching. Such mechanosensing allows *Xoo* to initiate effector delivery only after successful host engagement. Structural flexibility in HrpE is critical for these functions. Glycine-rich hinges allow localized bending, while conserved helical interfaces maintain overall filament stability. These features permit the pilus to withstand shear forces encountered in the apoplast and xylem while maintaining a continuous secretion pathway. Structural studies have demonstrated that this flexibility enables effectors to traverse the pilus lumen in an extended conformation without aggregation or blockage (Weber *et al.*, 2005). Collectively, the HrpE pilus is not merely a structural appendage but a multifunctional secretion organ that integrates mechanical resilience, signaling capacity, and transport efficiency, making it central to *Xoo* pathogenicity.

8.19.12 Translocon Assembly (HrpF)

The *hrpF* gene encodes the principal translocon protein responsible for forming the membrane-inserting pore at the tip of the T3SS pilus. This structure represents the final barrier between the bacterial secretion machinery and the host cytoplasm. Successful translocon assembly is therefore essential for productive effector delivery in *Xoo*.

One of the primary functions of HrpF is the creation of a localized discontinuity in the rice plasma membrane. Upon host contact, HrpF is secreted through the pilus and undergoes conformational changes that allow its insertion into the lipid bilayer. This insertion is tightly regulated to avoid excessive membrane damage that could trigger strong immune responses. Following insertion, HrpF oligomerizes to form a transmembrane channel through which effectors are directly delivered into the host cytosol. This channel bypasses the apoplastic space, preventing dilution or degradation of effectors and ensuring rapid intracellular delivery. Translocon assembly is also functionally coupled with pilus elongation. Proper alignment between the pilus tip and the host membrane is required for efficient pore formation. HrpF thus acts as a molecular interface linking bacterial secretion dynamics with host membrane penetration. The importance of HrpF is underscored by mutant analyses. In the absence of functional HrpF, effectors are still secreted but accumulate in the apoplast rather than entering host cells. This uncoupling of secretion and translocation leads to complete loss of virulence, highlighting that effector delivery, not secretion alone, is the critical determinant of pathogenicity. In addition, HrpF contributes to host specificity. Variations in translocon structure can influence compatibility with host membrane composition, suggesting that *Xoo* has optimized HrpF for interaction with rice plasma membranes.

8.19.13 Accessory Proteins (Hpa1, HpaB, HrpD6)

Accessory proteins encoded within or adjacent to the *hrp/hrc* cluster fine-tune the efficiency and timing of effector translocation. Although not structural components of the secretion apparatus, Hpa1, HpaB, and HrpD6 play critical regulatory and stabilizing roles. HpaB functions as a key effector chaperone and secretion regulator. By binding effectors in the cytoplasm, HpaB prevents premature folding and aggregation while also regulating secretion priority (Furutani *et al.*, 2017). This ensures that effectors are delivered in the correct temporal order once the T3SS becomes competent. Hpa1 is thought to enhance translocon efficiency and may function as a host contact facilitator. Its secretion prior to effectors suggests a role in priming the host membrane for translocon insertion. Hpa1 may also modulate host membrane properties, facilitating efficient pore formation. HrpD6 contributes to stabilization of assembly intermediates within the secretion apparatus. It is proposed to coordinate transitions between assembly stages, ensuring that pilus formation, translocon deployment, and effector secretion occur in a tightly regulated sequence.

Together, these accessory proteins increase secretion throughput, prevent mistimed effector release, and enhance the overall reliability of the T3SS during infection. Their roles highlight the importance of regulatory fine-tuning beyond the core structural machinery.

8.19.14 Integrative Role of *hrp/hrc* Systems in Pathogenicity

The pathogenic capacity of *Xoo* is inseparably linked to the function of the *hrp/hrc*-encoded T3SS. Rather than acting as isolated components, *hrp* and *hrc* genes form an integrated system that coordinates environmental sensing, structural assembly, and effector delivery. This integration allows *Xoo* to respond dynamically to host-derived signals, ensuring that energy-intensive secretion processes are activated only under favorable conditions. The *hrp/hrc* system thus acts as both a virulence apparatus and a regulatory hub. Disruption of any major *hrp/hrc* module compromises this integration, leading to failure of effector translocation and loss of pathogenicity. The system therefore represents the central pathogenicity island of *Xoo*.

8.19.15 Establishing Early Infection Competency

Early expression of *hrp* genes primes *Xoo* for infection before visible symptoms appear. This preparatory phase enables the bacterium to assemble a secretion-competent T3SS while still residing on the leaf surface or in intercellular spaces. By establishing infection competency early, *Xoo* can respond rapidly upon host contact, minimizing the delay between perception and effector delivery. This early readiness provides a competitive advantage during the initial stages of colonization.

8.19.16 Enabling Rapid Suppression of Defense Responses

Effector delivery through the T3SS allows *Xoo* to suppress plant immune responses at their earliest stages. Direct injection of effectors enables immediate interference with PTI signaling pathways before defense gene expression escalates. Effectors can also modulate ETI-related signaling components, preventing amplification of immune responses. In addition, certain effectors promote activation of host susceptibility genes, further enhancing pathogen fitness. In the absence of functional *hrp* genes, *Xoo* remains confined to the apoplast and triggers strong immune responses that severely limit bacterial growth.

8.19.17 Facilitating Colonization of Xylem Tissues

Although multiplication occurs primarily in the xylem, survival in the epidermis and mesophyll is a prerequisite for vascular entry. The T3SS aids early survival by suppressing local defenses and reducing physical barriers. Efficient effector translocation facilitates movement into vascular tissues, where *Xoo* encounters a distinct biochemical environment. Continued T3SS activity supports sustained colonization and systemic spread.

8.19.18 Determining Host Range

Variation within *hrp/hrc* genes plays a major role in determining host range and cultivar specificity. Structural differences in HrpE and HrpF influence pilus flexibility and translocon compatibility with host membranes. Accessory proteins such as HpaB further shape effector repertoires and secretion hierarchies, affecting interactions with host immune receptors. Even minor alterations in these components can shift compatibility outcomes. Thus, the Hrp/Hrc system not only enables pathogenicity but also defines the evolutionary boundaries of host adaptation in *Xoo*.

8.20 MULTILAYER REGULATORY ARCHITECTURE GOVERNING *HRP/HRC* GENE EXPRESSION IN *XOO*

The expression of the *hrp/hrc* gene cluster in *Xoo* is not constitutive; it is tightly controlled by an integrated regulatory network that ensures the T3SS is activated only under conditions resembling the plant apoplast. This multilayer regulation is essential because the T3SS imposes substantial metabolic costs, involves energetically expensive protein synthesis, and can expose the bacterium to host immune surveillance if expressed inappropriately (Wengelnik *et al.*, 1999; Jiang *et al.*, 2020). The regulatory system has evolved to synchronize effector translocation with environmental cues, host-derived signals, and intracellular checkpoints that collectively fine-tune the pathogenic program.

8.20.1 Environmental Sensing and Activation of the HrpG–HrpX Cascade

The primary switch controlling the Hrp/Hrc regulon is the hierarchical HrpG–HrpX cascade (Zou *et al.*, 2012). HrpG, an OmpR-family response regulator, functions as the master activator that integrates external conditions and relays the signal by inducing transcription of HrpX (Wengelnik *et al.*, 1999). Although the sensory kinase partner phosphorylating HrpG has not been definitively identified in *Xoo*, biochemical and genetic evidence suggests that its activation is linked to low nutrient availability, acidic pH, and the presence of plant-derived phenolics—conditions that mimic the rice apoplast.

Activated HrpX, an AraC-type transcription factor, binds conserved PIP (plant-inducible promoter) boxes upstream of *hrp/hrc* genes and effector genes, promoting transcription under host-like conditions (Furutani *et al.*, 2009). This ensures that effector production and secretion do not occur unless *Xoo* is in a plant-compatible environment. The tight dependency of each *hrp* operon on PIP motifs confers regulatory coherence across the cluster, enabling synchronized induction of structural components, secretion regulators, and effector substrates.

8.20.2 Secondary Environmental Inputs Modulating T3SS Expression

Beyond HrpG and HrpX, additional regulators fine-tune responsiveness to the rice environment:

- **Nutrient limitation signals**: Minimal media resembling apoplastic fluid elevate *hrp* expression by enhancing HrpG activity (Jiang *et al.*, 2020).
- **Iron limitation**: Rice apoplast iron scarcity enhances Hrp induction indirectly through iron-responsive regulators, reflecting the bacterium's perception of host colonization.
- **Osmotic and redox signals**: Environmental fluctuations characteristic of flooded paddy fields or oxidative bursts in rice tissue influence T3SS activation by modifying the expression of small RNAs and accessory regulators.

These inputs shape the temporal dynamics of *T3SS* gene expression during early colonization and xylem entry.

8.20.3 Small RNAs and Post-Transcriptional Regulation of the Hrp/Hrc Regulon

Post-transcriptional regulation adds an additional layer of specificity. Small RNAs (sRNAs) modulate the stability of HrpG and HrpX transcripts, ensuring rapid adjustments to changing conditions. In *Xoo*, several Hfq-dependent sRNAs influence the translation of T3SS-related mRNAs, although the complete set of regulatory RNAs remains under investigation. These sRNAs allow the pathogen to fine-tune expression without the delays associated with transcriptional adjustments, crucial during the early phases of rice infection.

8.20.4 Post-Translational and Protein-Level Control of T3SS Function

The protein assembly and function of the T3SS require gatekeeping mechanisms that coordinate effector secretion with apparatus assembly. HrcQ, HrcU, and Hpa proteins operate as checkpoints ensuring that:

- the basal body is assembled before pilus extension,
- early substrates (e.g., HrpE, translocon components) are secreted first,
- effector release begins only after the translocon has successfully engaged with the host membrane (Hartmann *et al.*, 2012).

Such post-translational control prevents premature effector leakage and ensures a sequential secretion hierarchy essential for efficient translocation.

8.20.5 Integration of Multiple Regulatory Layers during Rice Colonization

Collectively, transcriptional, post-transcriptional, and post-translational mechanisms form a regulatory architecture that enables *Xoo* to coordinate *hrp/hrc* gene expression with the biological context of infection. During initial leaf surface colonization, Hrp expression remains low; only upon sensing apoplastic conditions does the HrpG–HrpX cascade fully activate the system. As bacterial populations expand within the xylem, sRNA-mediated tuning and secretion hierarchy controls allow *Xoo* to maintain the T3SS while avoiding metabolic overload and excessive host immune recognition.

This multilayer regulation is central to the pathogen's ability to deploy the T3SS with high precision, ensuring that effector delivery is both timely and environmentally appropriate, ultimately supporting successful pathogenicity in rice.

8.21 TRANSCRIPTIONAL REGULATION OF *HRP* GENES IN *XOO* VIA *XYLR*

In *Xoo*, the *hrp* genes encoding T3SS are controlled by a two-component regulatory network: HrpG (an OmpR-type response regulator) activates HrpX, and HrpX (an AraC-type activator) drives transcription of the downstream *hrp/hrc/hpa* operons (Ikawa *et al.*, 2018). Importantly, the accumulation of HrpX protein depends strongly on the sugar environment: in minimal medium containing xylose, HrpX reaches higher levels than when other sugars (like glucose or fructose) are used, implying that xylose acts not only as a nutrient but as a signal that stabilizes HrpX by suppressing its proteolytic degradation (Ikawa and Tsuge, 2016).

Ikawa *et al.* (2018) showed that XylR, a LacI-type transcriptional repressor, is a central negative regulator: in a *xylR* deletion mutant, HrpX accumulates even without xylose, and *hrp* gene expression is elevated almost to the level of wild-type grown in xylose-containing medium. This strongly suggests that XylR serves as a molecular gatekeeper, preventing gratuitous T3SS activation under conditions lacking xylose.

Mechanistically, XylR does not appear to bind directly to the HrpX promoter. Instead, data indicate that XylR represses factor(s), likely proteases, that degrade HrpX in xylose-poor conditions. When xylose binds XylR, it induces a conformational change, weakening DNA binding, lifting repression, and allowing HrpX stabilization (Ikawa *et al.*, 2018).

8.21.1 *xylR* Operon and Its Regulation by Xylose

The *xylR* gene encodes a LacI-type repressor responsive to xylose. One of its principal targets is *xylA2*, one of two xylose isomerase genes in *Xoo* (the other being *xylA1*). Using DNA-binding assays, Ikawa *et al.* (2018) demonstrated that XylR binds upstream of *xylA2* (but not *xylA1*) and represses it in the absence of xylose. When xylose is present, it binds to XylR, diminishing its DNA-binding capacity and derepressing *xylA2*. This dual regulation links metabolic sensing to virulence: under xylose-rich conditions, *Xoo* increases xylose metabolism (via *xylA2*) and simultaneously primes its T3SS by stabilizing HrpX.

8.21.2 Xylose-Dependent Modulation: From Sensing to Virulence

In this system, xylose acts as a regulatory ligand rather than simply a nutrient. By binding to XylR, xylose relieves the repression on *xylA2* and on the (yet-unidentified) protease(s) that degrade HrpX. This derepression leads to increased xylose isomerase expression and HrpX stabilization, which in turn activates downstream *hrp* operons and promotes T3SS biogenesis (Ikawa *et al.*, 2018).

This mechanism constitutes a threshold-based switch: only when xylose concentration crosses a certain level—suggesting xylan breakdown in the host—does *Xoo* commit to T3SS expression. This helps the pathogen avoid unnecessary metabolic costs and reduces the risk of early host immune detection.

8.21.3 Significance of Xylose for *Xoo* Growth, Colonization, and Proliferation

Xylose functions in two critical ways for *Xoo*: as a *nutrient* and as a *signal* of host environment. The rice cell wall is rich in xylan, and its degradation (by either bacterial or plant enzymes) can liberate free xylose or xylo-oligomers into the apoplast. This xylose pool can serve as a cue that *Xoo* is in a favorable niche (Ikawa *et al.*, 2018).

When *Xoo* senses xylose via XylR, it adapts by upregulating *xylA2* (boosting its metabolic capacity) and by stabilizing HrpX (preparing its T3SS). This coupling ensures that virulence is expressed efficiently only when advantageous, thereby conserving energy and reducing premature immune activation (Ikawa and Tsuge, 2016).

In addition, *Xoo*'s regulatory network is more complex than just the XylR-HrpX axis:

- *GamR*, a LysR-type transcriptional regulator, directly activates both Hrp*G* and Hrp*X*, thus serving as a positive regulator of the T3SS regulatory cascade (Rashid *et al.*, 2016).
- The *MinCDE cell-division system* negatively regulates Hrp*G* and Hrp*X*, influencing T3SS expression, virulence, and motility (Yan *et al.*, 2022).
- Host-derived regulation also plays a role: overexpression of *OsNPR1* in rice suppresses expression of many *Xoo* virulence genes, including T3SS components (Dai *et al.*, 2023).

Thus, while the xylose–XylR circuit is central, it operates within a broader regulatory network that integrates metabolic cues, cell-cycle signals, and host factors.

8.21.4 Insights into BB Management from Xylose–XylR and GamR Regulation

The xylose–XylR–HrpX axis, together with GamR, offers promising targets for bacterial blight (BB) control:

1. **Chemical intervention:** Design molecules that mimic xylose but lock XylR into its repressive conformation, preventing T3SS activation.
2. **Genetic/breeding approaches:** Engineer rice to reduce xylan accessibility or apoplastic xylose levels, thereby disrupting the signal *Xoo* uses to trigger virulence.
3. **Synthetic biology/biocontrol:** Develop beneficial microbes that scavenge xylose or modulate its availability, acting as "signal sinks."
4. **Combined regulator targeting:** Target both metabolic sensing (XylR) and activators like GamR for a more robust disruption of T3SS.
5. **Integrated host resistance:** Enhance host immune pathways (e.g., via *OsNPR1*) to suppress pathogen virulence at the transcriptional level (Dai *et al.*, 2023).

8.21.5 Why Xylose and GamR Together Shape BB Development

The regulatory interplay between *XylR* and *GamR* gives *Xoo* a refined decision-making system: XylR senses host-derived xylose, controlling when to stabilize HrpX and commit to T3SS, while GamR ensures basal activation of the Hrp cascade under a variety of conditions by directly activating Hrp*G* and Hrp*X* (Rashid *et al.*, 2016). Meanwhile, other regulators (e.g., MinCDE, host factors) fine-tune this response based on cell division status and immune context. Disrupting any part of this multi-layered switch holds potential for innovative BB management.

9 Type III Secretion System Effectors

9.1 T3SS EFFECTOR REPERTOIRE OF *XOO*

Bacterial effectors secreted through the type III secretion system (T3SS), collectively referred to as T3SS-effectors (T3Es) (Table 9.1a,b), are central to the virulence strategy of *Xoo* (Jones and Dangl, 2006; Kay and Bonas, 2009; Kim *et al.*, 2008, 2009; Kumar and Mondal, 2013; Li *et al.*, 2018). The T3SS functions as a molecular syringe, delivering effectors directly into the host cytoplasm, thereby allowing the pathogen to manipulate host physiology and immune responses (Figure 9.1). T3Es are evolutionarily adapted to target crucial nodes of plant immunity, modulating both upstream perception mechanisms and downstream transcriptional programs.

9.1.1 Role of T3SS-effectors in Suppressing Pathogen-Triggered Immunity (PTI)

The primary barrier to pathogen colonization in plants is pathogen-triggered immunity (PTI), initiated upon recognition of pathogen-associated molecular patterns (PAMPs) or damage-associated molecular patterns (DAMPs) by cell-surface pattern recognition receptors (PRRs). Typical bacterial PAMPs include flagellin, lipopolysaccharides (LPS), and elongation factor Tu (EF-Tu), which activate signaling cascades involving mitogen-activated protein kinases (MAPKs), calcium influx, reactive oxygen species (ROS) production, and callose deposition at the cell wall (White *et al.*, 2009; Hu *et al.*, 2017; Ji *et al.*, 2018; Ishikawa *et al.*, 2014). T3Es interfere at multiple points in this pathway, suppressing PAMP recognition, disrupting signal transduction, and modulating transcriptional outputs, thus enabling bacterial proliferation within the host.

Recent transcriptomic analyses in rice have demonstrated that *Xoo* T3Es cause significant suppression of early PTI marker genes, including PR10, OsWRKY13, OsRLCK16, and OsFLS2, resulting in compromised basal immunity and enhanced susceptibility (Mondal *et al.*, 2016; Verma *et al.*, 2018; Ji *et al.*, 2022). Functional assays using deletion mutants of key effectors reveal that loss of individual T3Es leads to partial restoration of PTI responses, while multiple deletions produce additive effects, confirming the coordinated action of these effectors in evading host defenses.

DOI: 10.1201/9781003714590-9

TABLE 9.1A
Effector Repertoire of *Xoo* Strains with Locus IDs

Effector	PXO99A	KACC 10331	MAFF 311018	IXO_BB0003
xopAA	+ (PXO_00234)	+ (*XOO*3022)	+ (*XOO*_2875)	+
xopAB	+ (PXO_01867)	+ (*XOO*3383)	+ (*XOO*_3150)	+
xopAC	–	–	–	–
xopAD	+ (PXO_03833)	+ (*XOO*4401)	+ (*XOO*_4145)	+
xopAE	+ (PXO_03420)	+ (*XOO*0065)	+ (*XOO*_0110)	+
xopAF	–	–	–	–
xopAG	–	–	–	–
xopAH	? (PXO_2875)	–	? (*XOO*_3022)	?
avrBS1	–	–	–	–
avrBS2	+ (PXO_03330)	+ (*XOO*0168)	+ (*XOO*_0148)	+
xopB	+ (PXO_02108)	+ (*XOO*3424)	+ (*XOO*_3221)	+
xopC	+ (PXO_02107)	+ (*XOO*3426)	+ (*XOO*_3222)	+
xopD	–	–	–	–
xopE	–	–	–	–
xopF1	+ (PXO_03413)	+ (*XOO*0074)	+ (*XOO*_0103)	+
xopG	–	–	–	–
xopH	–	–	–	–
xopK	+ (PXO_01625)	+ (*XOO*1768)	+ (*XOO*_1169)	+
xopL	+ (PXO_01620)	+ (*XOO*1762)	+ (*XOO*_1162)	+
xopN	+ (PXO_02760)	+ (*XOO*0343)	+ (*XOO*_0315)	+
xopO	–	–	–	–
xopP	+ (PXO_02107)	+ (*XOO*3426)	+ (*XOO*_3222)	+
xopQ	+ (PXO_03901)	+ (*XOO*4466)	+ (*XOO*_4209)	+
xopR	+ (PXO_03819)	+ (*XOO*4391)	+ (*XOO*_4134)	+
xopT	+ (PXO_00236)	+ (*XOO*4824)	+ (*XOO*_2210)	+
xopU	+ (PXO_01625)	+ (*XOO*1768)	+ (*XOO*_1169)	+
xopV	+ (PXO_04172)	+ (*XOO*0433)	+ (*XOO*_3503)	+
xopW	+ (PXO_03356)	+ (*XOO*0134)	+ (*XOO*_0037)	+
xopX	+ (PXO_03702)	–	+ (*XOO*_4042)	+
xopY	+ (PXO_04865)	–	+ (*XOO*_1458)	+
xopZ1	+ (PXO_06125/01041)	+ (*XOO*2543)	+ (*XOO*_2402)	+
xopZ2	+ (PXO_06125/01041)	–	–	+

+: Effector present; In parenthesis, locus ID provided. –: Effector absent. ?: Presence uncertain; locus ID not confirmed.

TABLE 9.1B
***Xoo* Strains and Their TALEs No and Locus IDs**

Xoo Strains	No. of TALEs	TALE Locus ID (xanthomonas.org)
KACC10331	17	*XOO_*1132; *XOO_*1134; *XOO_*1136; *XOO_*1138; *XOO_*1996; *XOO_*1998; *XOO_*2001; *XOO_*2127; *XOO_*2129; *XOO_*2158; *XOO_*2160; *XOO_*2667;_Ψ *XOO_*2864; *XOO_*2865; *XOO_*2866; *XOO_*2868; *XOO_*4014
MAFF311018	17	*XOO_*1132; *XOO_*1134; *XOO_*1136; *XOO_*1138; *XOO_*1996; *XOO_*1998; *XOO_*2001; *XOO_*2127; *XOO_*2129; *XOO_*2158; *XOO_*2160; *XOO_*2667_Ψ; *XOO_*2864; *XOO_*2865; *XOO_*2866; *XOO_*2868; *XOO_*4014
PXO99a	19	PXO_03922 (tal1/pthXo7); PXO_00227 (tal2b/pthXo1); PXO_00223 (tal2a); PXO_00511_Ψ (tal3a); PXO_00505_Ψ (tal3b); PXO_00318 (tal4); PXO_00572 (tal5b/pthXo6); PXO_00567 (tal5a); PXO_00546 (tal6a); PXO_05609_Ψ (tal6b); PXO_05633 (tal7a); PXO_01085 (tal7b); PXO_06229 (tal8a); PXO_06234 (tal8b); PXO_02172 (tal9a); PXO_05714 (tal9b); PXO_05718 (tal9c/avrXa27); PXO_02269 (tal9d); PXO_02272 (tal9e)
Xoo race 4 (Indian strain)	15	tal1/pthXo7; tal2b/pthXo1; tal3a; tal4; tal5a; tal5b/pthXo6; tal6a; tal6b; tal7b; tal8a; tal8b; tal9a; tal9b; tal9c/avrXa27; tal9e

9.1.2 T3SS-effectors as Avirulence Factors and Effector-Triggered Immunity (ETI)

Beyond their role in virulence, certain T3SS-effectors function as avirulence factors when recognized by host immune receptors, particularly nucleotide-binding leucine-rich repeat (NLR) proteins (Hopkins *et al.*, 1992). Recognition triggers effector-triggered immunity (ETI), often associated with the hypersensitive response (HR), a form of programmed cell death that restricts pathogen spread (Popov *et al.*, 2018; Pruneda *et al.*, 2016). This dual functionality exemplifies the evolutionary arms race between *Xoo* and rice: effectors evolve to evade detection, while the host simultaneously evolves new NLRs capable of recognizing previously undetected effectors. Comparative studies demonstrate that ETI responses are highly specific and dependent on the precise amino acid composition of the effector and the configuration of the corresponding receptor (Yang and White, 2004).

9.1.3 Diversity of Effector Repertoires across *Xoo* Strains

Xoo exhibits remarkable diversity in its effector complement, shaped by geographic distribution, host selection pressures, and historical evolutionary events. This diversity underlies strain-specific virulence and host range, and represents a central feature in the evolutionary arms race between pathogen and rice cultivars. Comparative

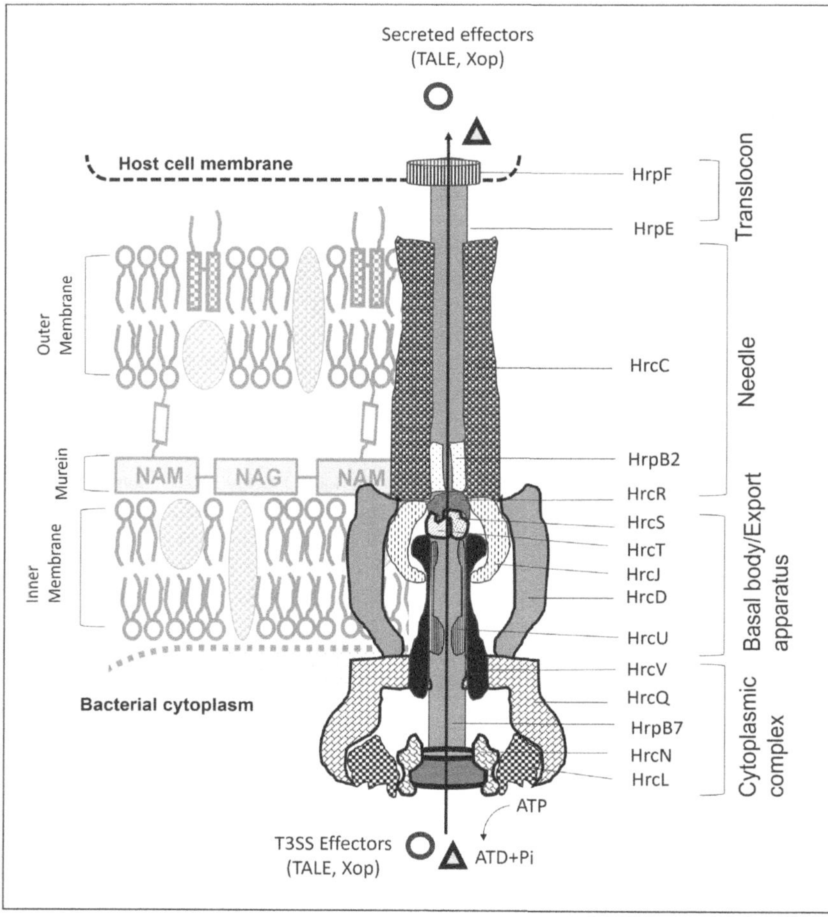

FIGURE 9.1 Schematic structure of bacterial type III secretion system (T3SS).

genomic analyses indicate that Indian strains, such as race 4, carry 21 Xop and 18 TALE effectors, whereas the Philippine PXO99A strain harbors 19 TALEs, and African strain K74 possesses 18 TALEs (Long *et al.*, 2018; Mondal *et al.*, 2014; Salzberg *et al.*, 2008; Yu *et al.*, 2015). Other sequenced strains such as KACC10331, MAFF311018, and PXO99A display variable numbers of TAL genes (6, 16, and 19, respectively) (Lee *et al.*, 2005; Ochiai *et al.*, 2005), reflecting regional evolutionary divergence.

This variation in effector repertoires is not random but represents an adaptive strategy for colonizing diverse rice hosts. Strains with a broader repertoire of TALEs or redundant Xops are able to target multiple host susceptibility genes simultaneously, enhancing virulence robustness. For example, Indian race 4 employs multiple

TALEs to activate SWEET11, SWEET13, and SWEET14, whereas PXO99A relies on a distinct combination of TALEs targeting overlapping or alternative EBEs. African strain K74, in contrast, appears to have evolved a more compact but efficient set of TALEs tailored to local rice varieties (Long *et al.*, 2018; Yu *et al.*, 2015).

Comparative genomics has revealed that variations in effector content arise from both gain and loss of effector genes as well as alterations in repeat-variable diresidue (RVD) composition within TALEs. In addition to gene count differences, sequence variations within CRRs modify DNA-binding specificity, thereby influencing which host genes are targeted. The diversity in TALE repertoires among *Xoo* strains demonstrates the pathogen's capacity to rapidly adapt to host genotypes, particularly in response to naturally occurring resistance alleles such as *xa13, xa25*, and *xa41* (Mondal *et al.*, 2014).

The genomic organization of effector genes also contributes to diversity. TALE genes are often clustered, whereas Xop effectors may be dispersed across the genome, often located within *hrp/hrc* islands or on plasmids, facilitating horizontal gene transfer. Lateral gene transfer, recombination, and gene duplication events have collectively driven diversification of the effector arsenal in *Xoo*, enabling it to overcome host defenses in geographically distinct regions (Mondal *et al.*, 2014; Kim *et al.*, 2008).

9.1.4 Functional Implications of Effector Diversity

The diversity of effectors directly impacts virulence and host range. Strains with multiple TALEs and Xops exhibit redundancy in targeting host susceptibility genes and defense pathways, increasing the robustness of infection. Redundant effectors can simultaneously activate distinct SWEET genes or suppress multiple nodes in the PTI cascade, ensuring successful colonization even if one target is mutated or blocked in the host (Long *et al.*, 2018; Mondal *et al.*, 2014; Hu *et al.*, 2017).

Furthermore, differences in effector repertoires explain the observed variability in lesion lengths, symptom severity, and bacterial multiplication rates among strains when infecting the same host cultivar. For instance, deletion mutants of individual TALE genes often exhibit reduced virulence, but strains with multiple redundant TALEs maintain high pathogenicity due to compensatory effects (Yang and White, 2004; Popov *et al.*, 2018; Pruneda *et al.*, 2016). Similarly, Xops targeting the plasma membrane, cytosol, or organelles provide complementary suppression of basal immunity, highlighting the multi-layered approach employed by *Xoo* to overcome plant defenses.

9.1.5 Molecular Drivers of Effector Variation

At the molecular level, effector diversification is influenced by both structural and regulatory mechanisms. TALEs exhibit modular architecture, where variations in CRRs and RVDs enable recognition of alternative EBEs, while Xops demonstrate conserved N- and C-terminal secretion signals coupled with highly variable central domains that encode enzymatic motifs and organelle-targeting signals. Such structural plasticity permits rapid evolution of novel virulence functions without disrupting the fundamental secretion and translocation machinery (White *et al.*, 2009; Singer *et al.*, 2013; Mondal and Kalaivanan, 2023).

Environmental and host-derived signals further shape effector expression. *hrp/hrc*-regulated genes respond to plant contact, pH, and nutrient availability, ensuring synchronized secretion of effectors during infection. Coupled with genomic variation, this regulatory plasticity allows *Xoo* populations to fine-tune effector deployment for maximal impact in a given host environment.

9.2 PHYLOGENETIC INSIGHTS AND EVOLUTIONARY PATTERNS

Phylogenetic analyses reveal that *Xoo* effectors cluster according to geographic origin and host adaptation. TALEs from Indian strains often form distinct clades separate from Philippine or African isolates, suggesting lineage-specific expansions. Similarly, Xops display diversification patterns that correlate with functional specialization and subcellular targeting preferences. Positive selection on RVD regions of TALEs and key catalytic motifs in Xops reflects ongoing co-evolution with host immune receptors (Mondal *et al.*, 2014; Kim *et al.*, 2008).

Horizontal gene transfer events, particularly involving plasmid-borne Xops, introduce novel effector functions into local populations, increasing genetic diversity and virulence potential. Recombination and gene duplication further expand the effector repertoire, creating combinatorial possibilities that enable rapid adaptation to evolving host resistance.

Effector diversity directly influences host compatibility. Strains with broad TALE repertoires can infect multiple rice cultivars, while those with limited effectors are restricted to susceptible hosts. Redundant and overlapping targeting of susceptibility genes ensures that single resistance alleles are often insufficient to block disease. Consequently, breeding for durable resistance requires knowledge of strain-specific effector repertoires and their cognate EBEs.

Natural resistance alleles such as *xa13*, *xa25*, and *xa41* demonstrate how mutations in EBE sequences can confer effective resistance. However, *Xoo* populations can evolve new TALE variants to recognize modified EBEs, suggesting the dynamic arms race between pathogen and host (Tran *et al.*, 2018; Oliva *et al.*, 2019). The interplay between effector diversity and host genotype shapes disease outcomes, complicating resistance deployment strategies in agricultural systems.

In summary, *Xoo* exhibits extraordinary diversity in TALE and Xop effectors, driven by evolutionary pressures, recombination, lateral gene transfer, and adaptation to host genotypes. Comparative genomics and functional analyses reveal that this diversity underpins virulence, host range, and strain-specific interactions with rice cultivars. Understanding these patterns is critical for predicting pathogen behavior, designing effective disease management strategies, and guiding the development of durable host resistance.

9.3 REGULATION OF EFFECTOR SECRETION

The expression and deployment of T3Es in *Xoo* are tightly regulated by the *hrp* (hypersensitive response and pathogenicity) and *hrc* (HR and conserved) gene clusters. The hrp regulon orchestrates transcriptional activation of effector genes in response to environmental cues such as pH, nutrient availability, and plant-derived signals, ensuring that effector secretion occurs primarily upon host contact (Alfano and Collmer, 2004; Espinosa and Alfano, 2004). In parallel, *hrc* genes encode structural components of the secretion apparatus and ATPase complexes, providing the energy for efficient translocation of effectors into the host cytoplasm. Disruption of *hrp* or *hrc* genes leads to complete loss of pathogenicity, underscoring the essential nature of this regulatory network.

9.4 FUNCTIONAL SPECIALIZATION OF EFFECTOR FAMILIES

T3Es in *Xoo* are broadly categorized into two groups based on their mode of action:

- **Transcription Activator-Like Effectors (TALEs):** These effectors enter the host nucleus and mimic eukaryotic transcription factors, binding to specific promoter sequences in susceptibility genes such as members of the SWEET family (Mondal *et al.*, 2021). Each TALE contains a conserved N-terminal secretion signal, a central repeat region (CRR) conferring DNA-binding specificity via repeat-variable di-residues (RVDs), nuclear localization signals (NLSs), and a C-terminal acidic activation domain (AD) (Figure 9.2). The modularity of TALEs allows the pathogen to evolve new DNA-binding specificities rapidly, targeting alternate susceptibility genes when host promoters are modified (Figure 9.3).
- **Xanthomonas Outer Proteins (Xops):** Non-TALE effectors exhibit high structural and functional diversity, targeting host signaling pathways, organelles, and enzymatic processes to suppress immunity. Approximately 40 multimembered families have been described, with conserved N- and C-terminal regions facilitating T3SS-mediated secretion and variable central domains dictating enzymatic function and subcellular localization (White *et al.*, 2009; Singer *et al.*, 2013). Functional studies of XopF, XopR, XopL, XopN, XopQ, and XopZ demonstrate their pivotal roles

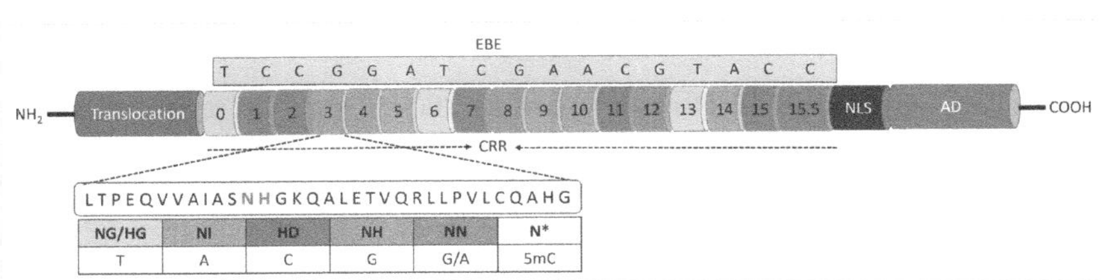

FIGURE 9.2 Structure of TALE protein showing a conserved N-terminal type III secretion signal for directing TALE into the host cell, a central repeat region (CRR) that binds specifically to its corresponding SWEET EBE. The multiple repeats have different RVDs, a nuclear localization signal (NLS) that permits TALE inside nucleus (black colored), and a C-terminal transcriptional acidic activation domain (AD) that activates TALE function (gray colored). The CRR region was enlarged, showing repeat variable di residues (RVD) at 12th and 13th positions. Table below indicates the specific recognition of nucleotides. (bottom row) by the RVDs (upper row). Effector binding element (EBE) with nucleotides, as shown in the shaded box in the CRR region, is based on the RVDs.

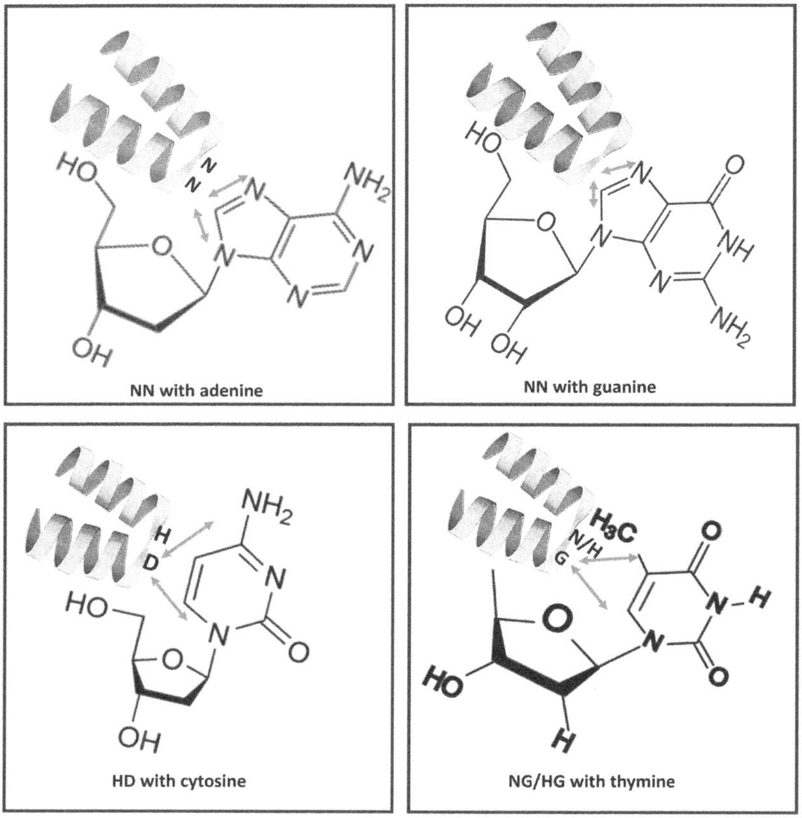

FIGURE 9.3 Binding affinity of TALE-RVDs with specific nucleotides.

in PTI suppression, PCD modulation, and host metabolic reprogramming (Akimoto-Tomiyama *et al.*, 2012; Deb *et al.*, 2019; Verma *et al.*, 2019; Ji *et al.*, 2022).

9.5 INTERACTIONS WITH HOST TARGETS

Xop effectors interact with a range of rice proteins to manipulate immune responses. For example, XopF interacts with cyclophilin and photosystem I subunit G (PSI-G), disrupting protein folding, ubiquitination, and photosystem function, thereby weakening basal immunity. XopR suppresses programmed cell death (PCD), enhancing bacterial proliferation and lesion development. Similar strategies have been observed in other host–pathogen systems, including XopD in tomato and XopJ in pepper (Kim *et al.*, 2008; Üstün *et al.*, 2013), highlighting the evolutionary conservation of immune suppression mechanisms across Xanthomonads.

Collectively, T3Es orchestrate a multifaceted assault on rice immunity, targeting PTI and ETI while reprogramming host metabolism to favor bacterial colonization. Comparative genomics, functional mutagenesis, and transcriptomic profiling reveal the dynamic interplay between effector diversity, regulatory control, and host susceptibility, illustrating the co-evolutionary arms race between *Xoo* and rice. Understanding the molecular specificity, regulation, and host targets of these effectors provides a foundation for designing durable resistance strategies and informs future breeding programs (Mondal *et al.*, 2014; Tran *et al.*, 2018; Oliva *et al.*, 2019).

9.6 TALE-DEPENDENT PATHOGENICITY OF *XOO* AND ITS REGULATION

Pathogenicity in *Xoo* is a highly orchestrated process, dependent on the precise interplay between the type III secretion system (T3SS), effector repertoire, and host-targeted regulatory strategies. TALEs serve as the central mediators of this process, converting host transcriptional machinery into a tool for bacterial proliferation (Alfano and Collmer, 2004; Espinosa and Alfano, 2004) (Figure 9.4).

9.6.1 Secretion and Delivery of TALEs

The T3SS, encoded by the *hrp* (hypersensitive response and pathogenicity) and *hrc* (HR and conserved) gene clusters, forms a needle-like apparatus spanning the bacterial envelope, enabling the direct injection of TALEs and other effectors into the host cytoplasm (Alfano and Collmer, 2004). The *hrp* genes primarily regulate transcription of effector genes and coordinate their timing, while *hrc* genes encode structural components and the ATPase complex that energizes effector translocation. Environmental cues such as pH, nutrient status, and plant surface signals precisely regulate this system to ensure that effectors are secreted only during host contact.

9.6.2 TALE–Host Promoter Interaction

Once inside the host nucleus, TALEs recognize specific effector-binding elements (EBEs) within target promoters (Table 9.2), particularly in the SWEET sugar transporter genes (Table 9.3). These transporters facilitate efflux of sucrose and other carbohydrates into the apoplast, providing a readily accessible nutrient source for the bacterium (Ji *et al.*, 2016a; Tran *et al.*, 2018; Oliva *et al.*, 2019). Well-characterized examples include the following:

- AvrXa7 and PthXa1 → SWEET11 promoter
- PthXa2 → SWEET13 promoter
- PthXa3, Tal5, TalC → SWEET14 promoter

The binding occurs through the central repeat region (CRR) of the TALE, where repeat-variable di-residues (RVDs) dictate DNA base specificity (Mak *et al.*, 2012, 2013). The C-terminal activation domain recruits host transcriptional machinery,

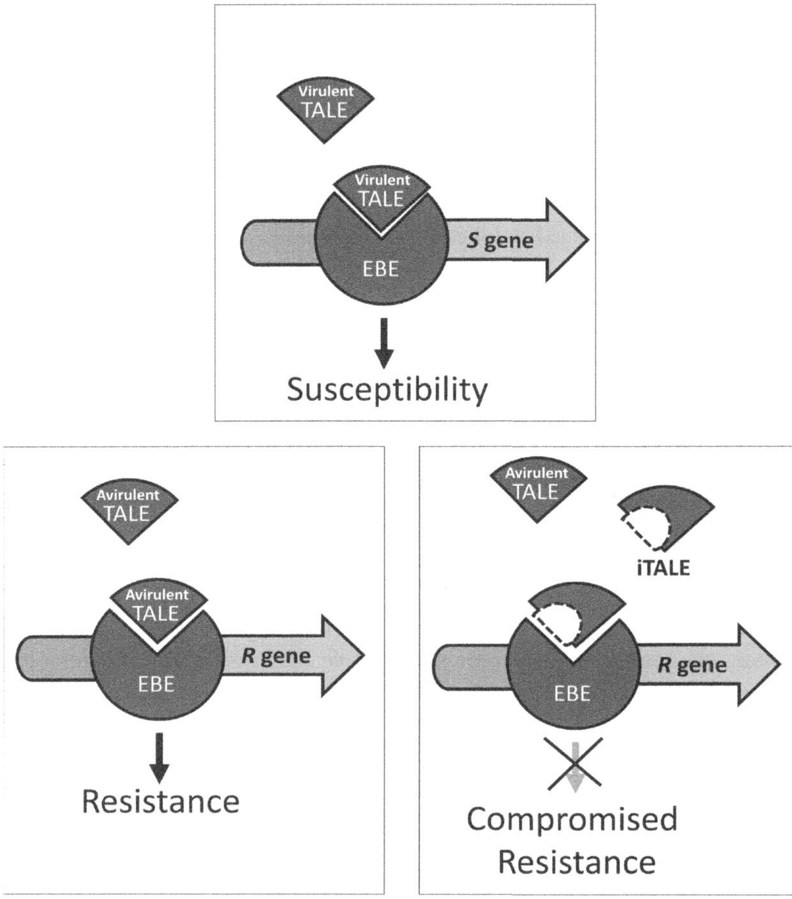

FIGURE 9.4 The schematic interaction between TALEs (virulent & avirulent types) and EBE region of the plant promoters for either *R* gene or *S* gene, leading to different phenotypes, namely susceptibility (upper one), resistance (lower left), and compromised resistance (lower right) as mediated by iTALE or truncated TALE.

enhancing mRNA synthesis from susceptibility genes and establishing an environment conducive to bacterial proliferation.

9.6.3 Redundancy and Virulence Robustness

The multiplicity of TALEs in each *Xoo* strain ensures redundancy in virulence. Some strains rely on a single dominant TALE to induce susceptibility, while others carry multiple TALEs targeting distinct SWEET promoters. This redundancy allows the

TABLE 9.2
Rice SWEETs Targeted by *Xoo* TALEs

SWEETs	SWEET EBEs in Rice (DNA)	*Xoo* TALEs	TALE RVDs (Protein)	Associated *R* Gene Allele
SWEET11	GCATCTCCCCCTACTGTACACCAC	Pthxo1	NN HD NI NG HD NG HD HD HD HD NG NG NI HD NG NI HD NI HD NI HD HD NI HD	*xa13*
SWEET13	ATAAAAGCACCACAACTCCCTT	PthXo2	NI HG NI NN NN NI NN HD NI HD NS NS NS HD NN HD NN HD NG HD HD HD NG NG	*xa25*
SWEET14	CATGCATGTCAGCAGCTGGTCAT	TalC	NS NG NS HD NI NG N NG HD NI NN N* NI NN HD NG NI NN N* HD NN NG	*Xa41*
	ATATAAACCCCCTCCAACCAGGTGCTAAG	PthXo3,	NI HG NI HG NI NI NI HD NN HD HD HD NG HD N* NI HD HD NN NS NI NN NN NG NN HD N*NS N*	
	ATAAACCCCCTCCAACCAGGTGCTAA	AvrXa7	NI HG NI NI NS HD NN HD HD HD NS N* N* HD HD NS NS NN NN NI NG NN NI N* NS N*	
	AAGCTCATCAAGCCTTCA	TalF	NN [HD/NN] [NN/NV] HD [NG/HD] HD NI [NI/NN] NN HD HD N* NG HD NI	

TABLE 9.3
Phylogenetic Clade of *SWEET* Genes

Clades	*SWEET* Genes	Sugar Transported	Functions
Clades I	*SWEET 1–3*	Hexose sugar (glucose and fructose)	-
Clade II	*SWEET 4–8*	Hexose sugar (glucose and fructose)	Pollen nutrition
Clade III	*SWEET 9–15*	Sucrose	Pollen loading, nectar secretion, seed filling, senescence
Clade IV	*SWEET 16–17*	(fructose)Vacuolar transporters and least frequent	Vacuolar storage of sugars

—Not defined.

pathogen to maintain high virulence even when host resistance alleles disrupt one or more EBEs. For example, strains containing both AvrXa7 and TalC can simultaneously activate *SWEET11* and *SWEET14*, amplifying carbohydrate availability and lesion development.

9.6.4 Co-Evolution with Host Resistance

TALE-dependent pathogenicity reflects an ongoing evolutionary arms race between *Xoo* and rice. Mutations in EBEs of SWEET promoters, such as those observed in resistance alleles *xa13, xa25*, and *xa41*, prevent TALE binding and confer recessive resistance (Hutin et al., 2015a; Haq *et al.*, 2022). In response, *Xoo* populations evolve new TALE variants with modified RVD arrangements capable of recognizing these altered EBEs, restoring virulence. This dynamic adaptation illustrates the plasticity of the TALE code and highlights the importance of understanding the modular architecture for resistance breeding.

Some isolates of *Xoo* produce shortened forms of transcription activator-like effectors known as interfering TALEs (iTALEs), which help them bypass plant resistance (Ji *et al.*, 2016b; Xu *et al.*, 2021, 2022a). Unlike normal TALEs, iTALEs do not have a transcription activation domain, but they still contain nuclear localization signals and are actively expressed, even though their genes were once thought to be non-functional (Xu *et al.*, 2022a).

The rice resistance gene *Xa1*, which encodes a nucleotide-binding leucine-rich repeat (NLR) protein, can recognize a wide range of TALEs and provide broad resistance against the pathogen (Ji *et al.*, 2020). However, the presence of iTALEs in many bacterial isolates disrupts this recognition, allowing the pathogen to overcome the resistance provided by *Xa1*. A variant type AvrXa23-like TALE enables *Xoo* to avoid being trapped by *Xa23* resistance gene in rice (Xu *et al.*, 2022b).

9.6.5 Regulatory Mechanisms of TALE Expression

TALE gene expression is tightly coordinated with the T3SS through the Hrp regulon. HrpX, an AraC-type transcription factor, acts as the master regulator, binding PIP box motifs upstream of effector genes to activate transcription. This synchronization ensures that TALE production aligns with secretion apparatus readiness and environmental cues, optimizing effector delivery during host colonization.

9.6.6 Functional Validation of TALE Activity

Functional analyses of TALE mutants reveal the additive and sometimes synergistic contributions of multiple effectors to virulence. Deletion of single TALE genes reduces lesion length and bacterial multiplication, while combined deletions dramatically attenuate pathogenicity. TALEs not only activate SWEET genes but also modulate host basal immunity by influencing hormone signaling and ROS production, highlighting their multifunctional role in infection (Mondal *et al.*, 2016; Verma *et al.*, 2019; Tran *et al.*, 2018).

9.6.7 Implications for Resistance Breeding

Understanding TALE–EBE interactions provides a framework for developing durable resistance strategies. By identifying critical EBEs and decoding TALE specificity, breeders can design rice varieties that either lack essential binding motifs or contain decoy sequences that sequester effectors without triggering susceptibility. Integration of structural, biochemical, and genetic analyses allows precise manipulation of host susceptibility loci to mitigate bacterial blight.

9.7 XOP EFFECTOR PROTEIN STRUCTURE

Xop effectors constitute a central component of the *Xoo* virulence arsenal, complementing the TALE family by manipulating host physiological and immune processes through diverse enzymatic and non-enzymatic mechanisms (Song and Yang, 2010; Mondal and Kalaivanan, 2023). Unlike TALEs, which act primarily in the nucleus, Xops are structurally and functionally heterogeneous and are deployed to multiple subcellular compartments, including the plasma membrane, cytoplasm, chloroplasts, and other organelles (Long *et al.*, 2018). This diversity enables precise modulation of plant immunity and host metabolism.

9.7.1 Classification and Structural Features

Approximately 40 multimembered Xop families have been characterized to date, each defined by highly conserved N- and C-terminal regions essential for type III secretion system (T3SS)-mediated translocation (White *et al.*, 2009; Singer *et al.*, 2013). These conserved domains act as secretion and translocation signals, ensuring proper recognition by the T3SS machinery. The central region of Xops is hypervariable, encoding domains responsible for enzymatic activity, subcellular targeting, and host specificity (Dean, 2011).

Common structural motifs found in Xops include a range of conserved protein domains that enable these effectors to directly manipulate host cellular processes. These motifs reflect convergent evolutionary strategies by which bacterial pathogens interfere with plant immunity, metabolism, and cellular organization. Although individual Xops differ in sequence and host targets, the presence of shared structural features highlights common principles underlying effector-mediated virulence.

9.7.1.1 Ubiquitin Ligase Domains

One of the most prominent structural motifs identified in several Xops is the ubiquitin ligase domain. These domains enable effectors to hijack the host ubiquitin–proteasome system, a central regulatory pathway controlling protein stability, signal transduction, and immune responses in plant cells. By functioning as E3 ubiquitin ligases or mimicking host ubiquitin ligase activity, Xops can tag specific host proteins for ubiquitination.

The ubiquitination of host immunity-related proteins often leads to their recognition and degradation by the 26S proteasome. Targets may include immune receptors, signaling kinases, transcription factors, or adaptor proteins involved in defense signaling. Through selective degradation of these components, Xops effectively dampen both basal immune responses and stronger resistance-associated pathways.

In addition to promoting protein degradation, ubiquitin ligase domains can alter protein localization or activity without complete destruction. Mono- or polyubiquitination can redirect host proteins away from signaling complexes or prevent their interaction with downstream partners. This allows Xops to fine-tune host responses rather than triggering broad cellular damage, which could alert immune surveillance systems.

The presence of ubiquitin ligase domains in Xops underscores the importance of post-translational regulation as a major target of bacterial virulence strategies. By manipulating a core host regulatory mechanism, these effectors exert wide-ranging effects on plant physiology while maintaining a relatively small molecular footprint.

9.7.1.2 Kinase or Phosphatase Motifs

Another frequently observed structural feature in Xops is the presence of kinase-like or phosphatase-like motifs. These domains allow effectors to interfere directly with phosphorylation-based signaling cascades, which are central to plant immune perception and response amplification.

Protein phosphorylation regulates the activation status, stability, and interactions of many immune-related proteins, including pattern-recognition receptors, mitogen-activated protein kinases (MAPKs), and transcriptional regulators. Xops containing kinase or phosphatase motifs can either introduce aberrant phosphorylation events or remove activating phosphate groups from key signaling components.

By acting as phosphatases, some Xops deactivate defense-associated kinases, effectively blocking downstream signaling. This suppresses the transcription of defense genes and limits the production of antimicrobial compounds. Conversely, kinase-like Xops may phosphorylate host proteins at non-native sites, disrupting normal signaling dynamics and creating non-functional or misregulated signaling states.

These phosphorylation-related activities allow Xops to interfere at very early stages of immune signaling, often before visible defense responses develop. Because phosphorylation is rapid and reversible, modulation of these pathways provides an efficient means for pathogens to control host responses in real time during infection.

Overall, kinase and phosphatase motifs highlight how Xops exploit the central role of phosphorylation in plant cellular regulation to suppress immunity and promote pathogen survival.

9.7.1.3 GTPase-Regulating Domains

Several Xops possess domains that regulate host small GTP-binding proteins, commonly referred to as small GTPases. These proteins function as molecular switches controlling diverse cellular processes, including vesicular trafficking, cytoskeleton organization, and membrane dynamics.

GTPases alternate between active (GTP-bound) and inactive (GDP-bound) states, and this cycling is tightly regulated by host factors. Xops with GTPase-regulating domains can disrupt this balance by accelerating GTP hydrolysis, preventing nucleotide exchange, or locking GTPases in inactive conformations.

Disruption of vesicular trafficking has profound consequences for plant immunity. Many defense-related proteins, including receptors and antimicrobial compounds, rely on vesicle-mediated transport for proper localization and secretion. By interfering with these pathways, Xops reduce the delivery of defense components to infection sites.

In addition, GTPase manipulation affects cytoskeletal dynamics, which are essential for cell wall reinforcement, intracellular transport, and immune signaling complex formation. Altered cytoskeleton organization can weaken physical barriers and impair signal transmission within the host cell.

The targeting of small GTPases demonstrates how Xops strategically interfere with host cellular logistics, ensuring that defense responses are spatially and temporally disrupted during infection.

9.7.1.4 Actin-Nucleation Domains

Actin-nucleation domains represent another important class of structural motifs found in Xops. These domains allow effectors to directly manipulate the host actin cytoskeleton, a dynamic network that supports cell shape, intracellular transport, and immune-related processes.

The actin cytoskeleton plays a critical role in plant defense by facilitating vesicle trafficking, receptor clustering, and targeted secretion of antimicrobial compounds. Xops with actin-nucleation activity can induce abnormal actin polymerization or depolymerization, leading to disorganized cytoskeletal structures.

By remodeling actin filaments, these effectors can inhibit the directed movement of defense vesicles toward infection sites. This reduces the accumulation of defense molecules at the bacterial entry point and compromises cell wall reinforcement. In some cases, cytoskeletal disruption may also interfere with signaling platforms required for immune activation.

Actin manipulation may additionally facilitate pathogen colonization by creating a more permissive intracellular environment. Altered cytoskeletal dynamics can affect cell rigidity and intercellular connectivity, potentially easing pathogen movement through host tissues.

The presence of actin-nucleation domains highlights the importance of cytoskeletal control as a virulence strategy and illustrates how Xops extend their influence beyond signaling pathways to fundamental aspects of cellular architecture.

9.7.2 Subcellular Localization and Targeting Signals

The effectiveness of Xops as virulence factors depends not only on their biochemical activity but also on their precise localization within host cells. Following delivery through the T3SS, Xops are guided to specific host compartments by intrinsic targeting signals embedded mainly within their central and C-terminal regions. These signals include short amino acid motifs, hydrophobic patches, and structurally defined domains that interact with host trafficking machinery. Such targeting ensures that each effector reaches the cellular site where its host substrates are located.

Experimental evidence from fluorescent protein fusion studies has demonstrated that Xops exhibit distinct and reproducible localization patterns within plant cells. For example, XopF predominantly accumulates at the plasma membrane, whereas XopR shows a more diffuse cytoplasmic or nuclear distribution. This spatial separation allows *Xoo* to manipulate multiple host defense layers in parallel rather than relying on a single point of interference. Compartmentalized effector action increases robustness of immune suppression and enables coordinated disruption of signaling, metabolism, and gene regulation.

Targeting specific subcellular locations also minimizes unintended cellular damage that could prematurely activate strong immune responses. By restricting activity to defined compartments, Xops exert focused and efficient control over host processes critical for defense.

9.7.2.1 Interference with PTI Signaling at the Plasma Membrane

The plasma membrane represents the first line of immune perception in plants. Pattern-recognition receptors and associated kinases are embedded in this membrane, where they detect pathogen-associated molecular patterns and initiate PTI signaling cascades. Xops that localize to the plasma membrane are therefore strategically positioned to interfere with immune activation at its earliest stage.

Membrane-targeted Xops often contain lipid-binding motifs or amphipathic helices that facilitate stable association with the inner leaflet of the plasma membrane. Once localized, these effectors can interact directly with receptor complexes, co-receptors, or downstream kinases. Such interactions may inhibit receptor phosphorylation, block complex assembly, or promote receptor degradation.

By suppressing PTI signaling at the membrane, *Xoo* prevents the rapid activation of downstream defense responses such as calcium influx, MAP kinase activation, and transcriptional reprogramming. This early interference delays immune recognition, allowing the pathogen to establish itself before stronger defense mechanisms are mobilized.

Plasma membrane localization also enables Xops to modulate membrane dynamics, including receptor endocytosis and recycling. Altered trafficking of immune receptors further weakens host surveillance and reduces sensitivity to pathogen detection.

9.7.2.2 Regulation of Reactive Oxygen Species Generation in the Cytoplasm

The cytoplasm is a central hub for immune signal transduction and metabolic integration. One of the earliest cytoplasmic defense responses is the production of reactive oxygen species, which function both as antimicrobial agents and as secondary messengers in immune signaling.

Cytoplasm-localized Xops are able to interfere directly with ROS production and signaling pathways. These effectors may interact with enzymes responsible for ROS generation or with regulatory proteins that control their activation. By dampening ROS accumulation, Xops reduce cellular stress signals that would otherwise amplify immune responses.

In addition to limiting ROS production, cytoplasmic Xops may interfere with ROS-mediated signaling by altering redox-sensitive proteins. Many immune regulators rely on controlled redox changes for activation. Disruption of this balance compromises downstream signaling and weakens defense gene induction.

Cytoplasmic localization also allows Xops to influence other defense-related processes, such as protein phosphorylation cascades and hormone signaling pathways. Through coordinated interference, these effectors reduce the effectiveness of cytoplasmic immune integration and signal amplification.

9.7.2.3 Manipulation of Transcriptional Regulation in the Nucleus

The nucleus is the control center for immune gene expression. Following immune activation, defense signals converge on the nucleus, leading to large-scale transcriptional reprogramming. Xops that localize to the nucleus are therefore able to exert profound effects on host defense outcomes.

Nuclear-targeted Xops often contain nuclear localization signals that facilitate their import through nuclear pore complexes. Once inside the nucleus, these effectors can interact with transcription factors, chromatin modifiers, or components of the transcriptional machinery.

By inhibiting defense-related transcription factors or altering chromatin accessibility, nuclear Xops suppress the expression of immune genes. This includes genes involved in antimicrobial compound production, cell wall reinforcement, and systemic signaling. Some effectors may also promote expression of susceptibility-associated genes, further enhancing pathogen fitness.

Nuclear localization enables sustained immune suppression because transcriptional effects persist even after initial signaling events subside. This allows *Xoo* to maintain a favorable intracellular environment throughout infection progression.

9.7.2.4 Metabolic Modulation in Chloroplasts and Mitochondria

Chloroplasts and mitochondria play essential roles in energy production, redox balance, and immune signaling. These organelles are also major sources of defense-associated metabolites and signaling molecules. Targeting of Xops to these compartments allows *Xoo* to influence host metabolism and stress responses at a fundamental level.

Chloroplast-localized Xops can interfere with photosynthesis-related processes that contribute to defense signaling. Chloroplasts are involved in the synthesis of defense hormones and ROS, and disruption of these pathways weakens immune responses. Effectors targeting chloroplasts may alter metabolite fluxes, reducing the availability of energy and precursors needed for defense.

Mitochondrial targeting enables Xops to modulate respiration and programmed cell death pathways. By stabilizing mitochondrial function or suppressing stress-induced signaling, these effectors prevent premature cell death that could limit pathogen spread.

Metabolic modulation also allows the pathogen to redirect host resources toward its own growth. By reshaping energy allocation and redox balance, Xops create conditions that favor bacterial multiplication while maintaining host cell viability.

9.7.2.5 Integrated Significance of Compartmentalized Effector Targeting

The ability of Xops to localize to multiple subcellular compartments underscores a sophisticated virulence strategy based on spatial precision. Rather than relying on a single mode of action, *Xoo* deploys effectors that collectively suppress immunity at the membrane, cytoplasmic, nuclear, and organellar levels.

This compartmentalization enables simultaneous disruption of perception, signaling, transcription, and metabolism. Such multi-layered interference ensures robustness of immune suppression and reduces the likelihood that host defenses can compensate through redundant pathways.

Subcellular localization and targeting signals are central determinants of Xop function. By directing effectors to specific host compartments, *Xoo* maximizes the impact of its virulence arsenal and achieves efficient, sustained colonization of rice tissues.

9.7.3 Functional Validation of Xop Effectors

Key Xop effectors in *Xoo*, including XopF, XopR, XopL, XopN, XopQ, and XopZ, have been experimentally validated for their roles in virulence. Hrp-dependent expression of these effectors and their translocation into rice cells have been confirmed using adenylate cyclase reporter assays (Akimoto-Tomiyama *et al.*, 2012; Deb *et al.*, 2019; Verma *et al.*, 2018, 2019; Ma *et al.*, 2020; Ji *et al.*, 2022).

Deletion mutants of *ΔxopF* and *ΔxopR* exhibit significantly reduced in planta growth, attenuated lesion formation, and failure to suppress PAMP-triggered immunity (PTI), highlighting their critical role in establishing infection (Figures 9.5 and 9.6). Expression analyses indicate upregulation of PTI marker genes, such as *PR10*, *OsWRKY13*, *OsRLCK16*, and *OsFLS2*, in mutant-infected rice plants, confirming their immune-suppressive function. Enhanced callose deposition in these mutants further underscores the importance of Xops in circumventing basal defense (Mondal *et al.*, 2016; Verma *et al.*, 2018, 2019).

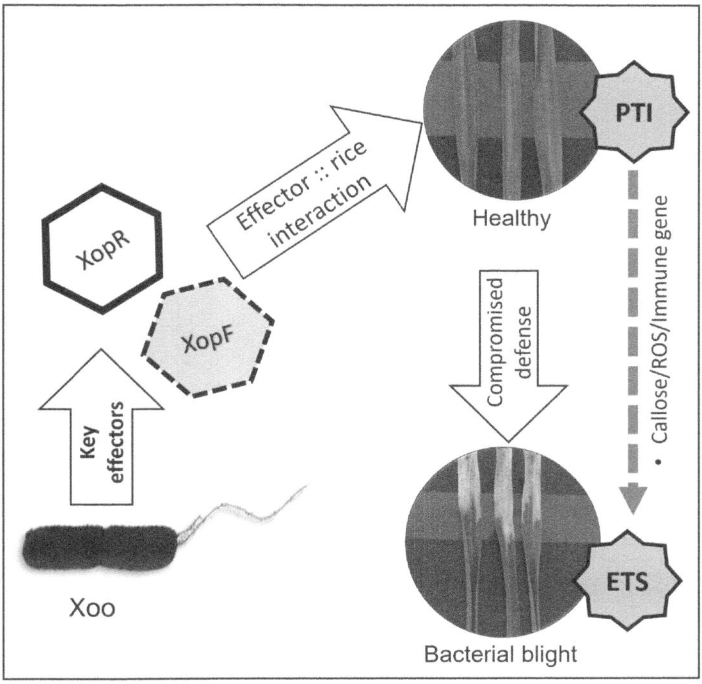

FIGURE 9.5 Modulation of rice immune system by two key Xop effectors of *Xoo*.

9.7.4 Protein–Protein Interactions

Xop effectors manipulate host physiology through direct protein–protein interactions. For instance, XopF interacts with cyclophilin and photosystem I subunit G (PSI-G), disrupting protein folding, ubiquitination, and photosystem activity, thereby compromising host defenses. Network analyses reveal extensive connectivity of these host interactors with other components of defense and metabolic pathways, highlighting the multi-targeted impact of Xop effectors (Figure 9.7).

Structural docking using Phyre2 homology models and HADDOCK simulations indicate that hydrogen bonding is a major stabilizing force in these interactions, providing a molecular basis for effector function (Figure 9.8).

9.7.5 Suppression of Programmed Cell Death (PCD)

XopR has been shown to mitigate PCD in rice, a defense response often triggered during effector recognition. Trypan blue staining of *ΔxopR or hrpX* mutants demonstrates extensive cell death zones, whereas complementation restores wild-type suppression. Ectopic expression of XopR in Nicotiana benthamiana similarly reduces visible PCD symptoms, and ion leakage assays confirm enhanced cell integrity in the

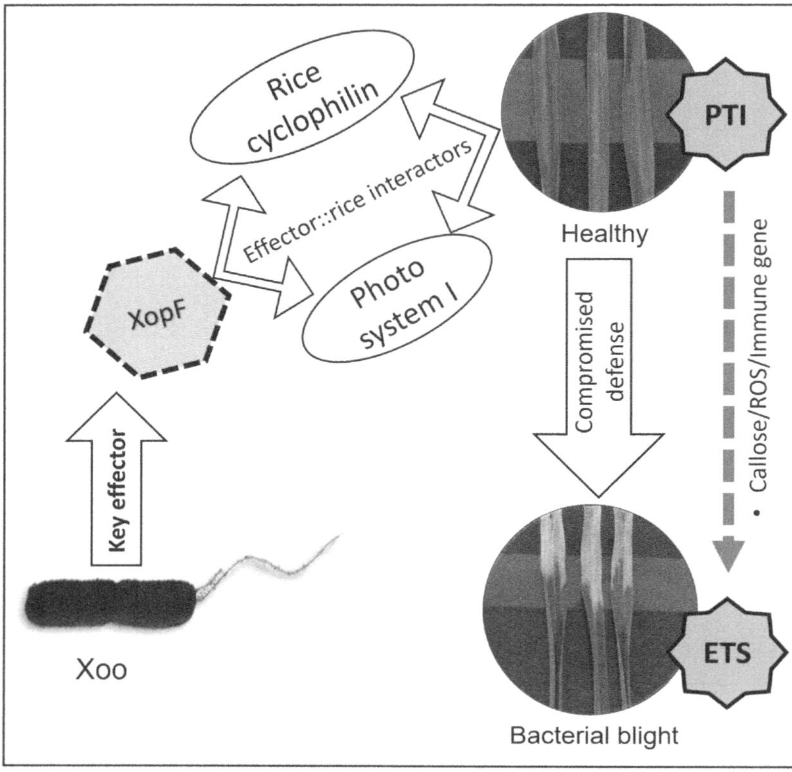

FIGURE 9.6 Schematic outline showing underlying interaction between XopF, the T3SS effector of *Xoo*, and rice proteins leading to modulation of PTI to ETS via compromised plant defense.

presence of the effector. This suppression allows *Xoo* to maintain tissue colonization and lesion expansion.

9.7.6 Comparative Analysis across *Xanthomonas* Species

Many Xop effectors exhibit conservation across Xanthomonas species, suggesting an evolutionarily conserved virulence mechanism. For example:

- XopJ (*X. euvesicatoria*) and XopF (Indian *Xoo*) both suppress PTI at the plasma membrane (Üstün *et al.*, 2013; Kumar *et al.*, 2016).
- XopD in tomato shares functional similarity with XopR in rice, targeting MAMP-triggered immune responses (Kim *et al.*, 2008).

Such conservation underscores the strategic importance of Xop-mediated immune suppression in host–pathogen co-evolution.

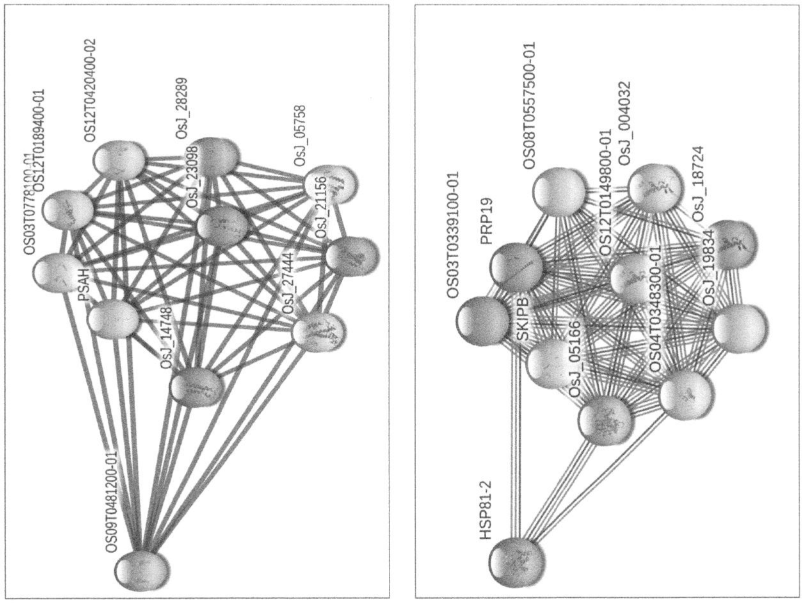

FIGURE 9.7 STRING-based networking of the two rice interactors, PSI-G protein (left) and cyclophilin II (right).

9.7.7 Transcriptomic Impact on Rice Host

Global transcriptomic profiling of rice infected with virulent *Xoo* strains reveals that Xop effectors extensively remodel host gene expression. Early and late defense-responsive genes, including peroxidases, NLRs, cytochrome P450s, and pathogenesis-related proteins, are dynamically modulated. Key enzymatic categories such as kinases, transferases, and synthetases exhibit pronounced alterations, indicating broad metabolic reprogramming during infection (Mondal *et al.*, 2022a; Amrutha *et al.*, 2023; Kong *et al.*, 2020). The analyses revealed substantial modulation of transcripts associated with signal perception, transcriptional regulation, and defense expression during bacterial blight development. Temporal expression profiling identified both early and late defense-responsive genes, including *peroxidases*, *NLRs*, *cytochrome P450s*, and *pathogenesis-related* proteins, which were dynamically regulated. Key enzymatic categories such as kinases, transferases, and synthases/synthetases exhibited pronounced alterations (Kong *et al.*, 2020), suggesting broad metabolic reprogramming of rice cells during infection.

Altogether, these studies emphasize the intricate molecular interplay between *Xoo* and rice. Xop effectors orchestrate multifaceted suppression of host defenses and redirect host metabolic processes to favor bacterial virulence, while rice counteracts through dynamic transcriptional reprogramming. Identifying and functionally validating rice targets of these effectors will deepen our understanding of host susceptibility and pave the way for breeding or engineering durable resistance to bacterial

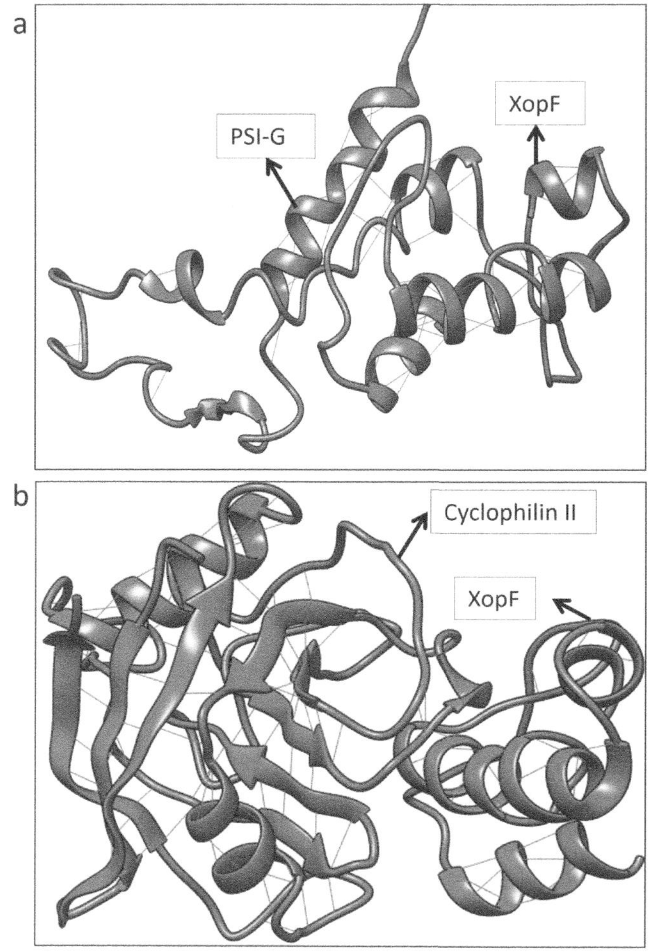

FIGURE 9.8 Protein docking between XopF and interactors PSI-G (a) and cyclophilin II (b). The models were generated using the Phyre2 server under normal modeling mode. Hypothetical models were constructed through homology modeling, and high-ambiguity driven protein–protein DOCKing (HADDOCK) was employed to simulate interactions, visualized using BioVia Discovery Studio. Light lines connecting the model chains indicate hydrogen bonds, which are critical for protein–protein interaction stability.

blight. However, the molecular roles of many Xop effectors remain to be clarified through future wet-laboratory investigations.

Xop effectors in *Xoo* are versatile virulence factors that manipulate multiple host processes to favor infection. Their structural diversity, subcellular targeting, biochemical activities, and regulatory control underpin the pathogen's success. Functional validation, interaction network analyses, and transcriptomic profiling collectively highlight the sophisticated strategies employed by *Xoo* to suppress immunity, control PCD, and reprogram host metabolism. Future studies focused on uncharacterized Xops will further illuminate the molecular basis of host susceptibility and inform strategies for durable bacterial blight resistance.

10 Genomic Features of *Xoo*

Understanding the genomic architecture of *Xoo* is central to elucidating its pathogenicity, adaptability, and co-evolution with rice hosts. The genomic features of *Xoo* reveal a pathogen with a highly conserved core genome coupled with a remarkable capacity for structural variability and adaptation. Differences in genome size, TAL effector repertoires, IS element load, regulatory alleles, and accessory gene content collectively shape the regional pathogenic potential of individual strains. Comparative genomics of Asian and African isolates emphasizes the dynamic interplay between conserved cellular functions and variable virulence determinants. Genome plasticity, facilitated by insertion sequences, plasmids, prophages, horizontal gene transfer, and CRISPR-Cas–mediated immunity, enables *Xoo* to rapidly respond to host resistance pressures and environmental changes. This genomic flexibility underpins the emergence of novel pathotypes, necessitating continuous monitoring, functional characterization, and integration of genomic insights into rice breeding and disease management strategies. A deep understanding of these genomic features is essential for predicting the evolution of bacterial blight, guiding the deployment of resistant rice cultivars, and designing sustainable control measures. The combination of detailed genome annotation, comparative analyses, and functional genomics offers a comprehensive framework for dissecting *Xoo* biology and mitigating its impact on global rice production.

Genome sequencing of multiple *Xoo* strains has revealed a combination of highly conserved core regions and highly variable genomic islands, reflecting a balance between essential cellular function and evolutionary adaptation. The first complete genome of *Xoo*, the Philippine strain PXO99A, unveiled a circular chromosome of approximately 5.24 megabase pairs (Mbp) with a G+C content of 63%, encoding over 5,000 protein-coding genes (Mondal *et al.*, 2020c). Subsequent sequencing of Korean, Japanese, Indian, and African strains has expanded our understanding of genome diversity within this species, highlighting strain-specific differences in virulence, host specificity, and structural plasticity.

The genome of the Indian strain IXOBB0003, recently sequenced from the northwestern rice-growing regions, has been deposited in NCBI GenBank under accession number CP098032. Comparative MAUVE alignments of IXOBB0003 with PXO99A (Philippines), KACC10331 (Korea), MAFF311018 (Japan), and BAI3 (Africa) reveal highly conserved core regions punctuated by genomic rearrangements, insertions, deletions, and novel effector loci (Table 10.1). These findings underscore the role of genome plasticity in the emergence of new *Xoo* pathotypes adapted to local rice cultivars (Kalaivanan *et al.*, 2023).

 DOI: 10.1201/9781003714590-10

TABLE 10.1
Comparative Insights across Genomic Regions of *Xoo* Strains

Region/Strain Example	Genome Size (Mbp)	Genes (~)	TAL Effector Repertoire	IS Elements	Distinctive Features
Philippines (PXO99A)	5.24	5,044	>18 (largest known)	>300	Reference TAL effector-rich genome; highly virulent
Korea (KACC10331)	4.94	4,600	~9–10	~250	Smaller genome; well-characterized CRISPR-Cas system
Japan (MAFF 311018)	5.05	~4,800	12–14	~260	Moderate TAL diversity; stable genomic structure
India (IXO704)	5.13	~5,000	12–16	>280	High IS density; rapid emergence of new pathotypes
China (GX01)	5.12	4,950–5,050	10–14	>270	Distinct TAL variants; allelic variation in *rpf* genes
India (IXO_BB0003)	5.08	~4,980	~13	~280	Plasmid-encoded factors; unique TAL-target patterns linked to Indian cultivars

10.1 GENERAL GENOME CHARACTERISTICS

Xoo genomes typically comprise a single circular chromosome, with size variations reflecting strain-specific expansions or contractions of genomic regions. The Philippine strain PXO99A is the largest among sequenced strains at approximately 5.24 Mbp, while the Korean isolate KACC10331 is relatively smaller at 4.94 Mbp. Japanese MAFF311018 and Indian IXO704 strains have intermediate sizes of 5.05 Mbp and 5.13 Mbp, respectively, whereas IXOBB0003 measures 5.08 Mbp. Despite this size variation, gene content scales proportionally with genome length, with PXO99A encoding roughly 5,044 genes, KACC10331 around 4,600, MAFF311018 about 4,800, IXO704 approximately 5,000, and IXOBB0003 encoding 4,980 protein-coding genes.

The G+C content remains relatively stable across all strains, ranging from 63% to 64%, consistent with other members of the *Xanthomonas* genus. Most isolates lack extrachromosomal plasmids; however, certain Indian strains, including IXOBB0003, harbor small plasmids carrying accessory genes that enhance virulence, environmental fitness, or metabolic flexibility. The high coding density and limited intergenic space highlight the evolutionary streamlining of *Xoo* genomes, favoring functional versatility while minimizing non-essential DNA.

Structural comparisons across strains reveal that while the core genome maintains essential housekeeping functions such as DNA replication, transcription, translation, and energy metabolism, accessory regions are highly variable. These include loci encoding TAL effectors, type III secretion system effectors, extracellular enzymes, and polysaccharide biosynthesis genes, all of which influence virulence and host specificity.

10.2 VIRULENCE AND PATHOGENICITY DETERMINANTS

The pathogenic success of *Xoo* hinges upon a repertoire of virulence genes that mediate host colonization, suppression of immunity, and nutrient acquisition. Among these, transcription activator-like effectors (TALs) constitute the most variable and functionally critical component. TAL effectors are modular proteins that enter host nuclei via the type III secretion system, bind specific DNA sequences in host promoters, and activate susceptibility genes to facilitate infection. PXO99A possesses more than 18 TAL effectors, whereas KACC10331 has a smaller set of approximately 9–10. MAFF311018 encodes 12–14 TALs, while Indian strains IXO704 and IXOBB0003 carry 12–16 effectors, including novel allelic variants not observed in other regional isolates. The unique combinations of TAL effectors in IXOBB0003 reflect adaptation to the genetic makeup of Indian rice cultivars, underscoring the role of regional host selection in shaping effector diversity.

The type III secretion system is highly conserved across *Xoo* strains, comprising a core set of *hrp* and *hrc* genes that assemble a syringe-like apparatus to translocate both TAL and non-TAL effectors into host cells. This system is critical for establishing infection, suppressing host immunity, and modulating host physiology.

Exopolysaccharide production, mediated by the gum operon, is another conserved virulence determinant. EPS facilitates bacterial adhesion to host surfaces, protects cells against environmental stresses, and promotes colonization of the xylem. Indian strains such as IXOBB0003 display robust EPS production, enhancing their ability to invade and spread through rice vascular tissue.

Xoo also produces a suite of extracellular enzymes, including cellulases, xylanases, and proteases, which degrade plant cell walls and facilitate nutrient acquisition. Xylanase activity in Indian isolates is particularly pronounced, consistent with the aggressive colonization observed in local rice varieties. The regulation of these enzymes, along with EPS synthesis, is orchestrated by the rpf cluster, which exhibits allelic diversity across Chinese and Indian isolates, modulating quorum-sensing–dependent pathogenic responses.

10.3 GENOMIC PLASTICITY AND STRUCTURAL VARIATION

Genomic plasticity is a hallmark of *Xoo* evolution and a key factor in the emergence of novel pathotypes. Insertion sequences and transposable elements are abundant across all strains, with PXO99A containing over 300 IS elements and IXOBB0003 approximately 280. These elements drive chromosomal rearrangements, duplications, and inversions, contributing to the structural variability observed among regional strains. Comparative analyses have revealed multiple inversions and large-scale duplications in Indian and Chinese isolates, highlighting the dynamic nature of *Xoo* genomes.

The CRISPR-Cas system is present in several *Xoo* strains, including KACC10331 and IXOBB0003, and demonstrates highly strain-specific spacer content. These spacers provide a historical record of bacteriophage interactions and suggest that phage-mediated selection pressure has shaped genome evolution differently in each region.

Plasmids and prophages further contribute to genomic diversity. Some Indian strains harbor plasmids that carry accessory virulence genes, while multiple strains exhibit prophage remnants indicative of historical phage integration. Such mobile elements serve as reservoirs of genetic variation that can be co-opted for adaptation, increasing the capacity for rapid evolution in response to environmental or host-imposed selective pressures.

10.4 COMPARATIVE GENOMICS ACROSS ASIAN AND AFRICAN STRAINS

Comparative genomic studies across Asian and African *Xoo* strains reveal a conserved core genome responsible for fundamental cellular processes and variable regions that mediate pathogenicity and host adaptation. PXO99A, KACC10331, MAFF311018, IXO704, IXOBB0003, and BAI3 share a substantial fraction of genes related to replication, transcription, translation, and primary metabolism, forming a stable genomic backbone. In contrast, TAL effector repertoires, rpf alleles, and insertion sequence distribution are highly strain-specific, reflecting adaptation to local rice cultivars and ecological conditions.

African strains such as BAI3 display a reduced TAL effector complement compared to Asian isolates but harbor unique virulence determinants that facilitate infection of African rice varieties. Structural variations, including inversions, duplications, and mobile element insertions, differ between African and Asian strains, suggesting independent evolutionary trajectories influenced by regional selective pressures. These genomic distinctions provide insight into the emergence of geographically restricted pathotypes and highlight the role of genome plasticity in shaping pathogenic diversity.

10.5 REGIONAL ADAPTATION AND HOST–PATHOGEN CO-EVOLUTION

The diversity of TAL effector repertoires among *Xoo* strains is a primary determinant of regional adaptation and host specificity. TAL effectors act as molecular keys that manipulate host gene expression to favor infection. In India, the IXOBB0003 strain carries TALs targeting promoters of susceptibility genes specifically present in widely cultivated rice varieties, resulting in high virulence in local cultivars. Similarly, the Indian IXO704 strain exhibits TAL variants not observed in PXO99A or KACC10331, suggesting independent evolution in response to rice cultivar diversity.

Japanese and Korean strains exhibit different TAL combinations, reflecting the genetic structure of rice varieties in East Asia. For example, MAFF311018 targets a subset of rice susceptibility genes distinct from PXO99A, while KACC10331 has a reduced set of TALs, correlating with the presence of partially resistant rice cultivars in Korea. African strains such as BAI3 have evolved independently, showing a reduced TAL complement but unique effector combinations that enable infection of local *Oryza glaberrima* and hybrid varieties.

Beyond TAL effectors, regional adaptation is also reflected in other virulence determinants. Variations in rpf alleles, exopolysaccharide biosynthesis genes, and extracellular enzyme expression modulate bacterial colonization and disease progression. The interplay between these determinants and host resistance genes drives a dynamic co-evolutionary arms race, resulting in the continuous emergence of novel pathotypes in geographically distinct rice-growing regions.

Structural genome variations such as chromosomal inversions and IS-mediated rearrangements also contribute to adaptation. For example, inversions observed in Indian isolates often coincide with TAL effector loci, potentially facilitating the duplication of effectors and rapid generation of novel allelic variants. The co-occurrence of prophage sequences and plasmids in some isolates further introduces new virulence genes or regulatory elements, enhancing regional fitness.

10.6 MOLECULAR MECHANISMS GOVERNING GENOME VARIABILITY

The plasticity of *Xoo* genomes is underpinned by multiple molecular mechanisms that collectively enhance adaptability and pathogenic potential. Insertion sequences (IS elements) are a major driver of genome rearrangements. Their high abundance, ranging from approximately 280 in IXOBB0003 to over 300 in PXO99A, mediates duplications, deletions, and inversions that can alter gene dosage or expression patterns. These elements frequently flank TAL effector clusters and other virulence loci, suggesting a mechanistic basis for effector diversification.

Horizontal gene transfer, facilitated by plasmids and phage-mediated transduction, introduces novel genes into *Xoo* genomes. Certain Indian isolates harbor plasmids carrying accessory virulence genes or regulators, which may confer selective advantages under specific environmental or host-imposed pressures. Prophage remnants embedded within chromosomes provide additional sources of genetic variation, occasionally contributing functional genes that modulate pathogenicity or stress responses.

The CRISPR-Cas system contributes to genome stability while simultaneously recording the history of bacteriophage interactions. Spacer sequences in strains such as IXOBB0003 are highly strain-specific, indicating exposure to diverse phages in different ecological niches. These adaptive immune elements not only prevent deleterious phage integration but also influence the acquisition of mobile genetic elements, indirectly shaping genomic architecture.

Collectively, these mechanisms create a dynamic equilibrium in which the *Xoo* genome maintains core functions while allowing extensive variability in regions responsible for host interaction, virulence, and environmental adaptation. This balance underpins the emergence of new pathotypes capable of overcoming regional rice resistance genes.

10.7 FUNCTIONAL GENOMICS AND APPLICATIONS

Insights from *Xoo* genomics have practical applications in breeding and disease management. Understanding TAL effector diversity enables the identification of susceptibility genes targeted in specific rice cultivars. This knowledge facilitates the development of rice lines carrying modified promoters or resistant alleles, thereby conferring durable resistance. Functional genomics approaches, including transcriptomic profiling of infected plants and knockout studies in bacterial strains, further elucidate the roles of specific effectors, regulatory genes, and mobile elements.

Genome-wide association studies in *Xoo* can reveal correlations between specific virulence factors and host susceptibility patterns, guiding targeted breeding programs. The identification of conserved virulence determinants such as the T3SS, gum operon, and extracellular enzymes provides potential targets for chemical or biological control strategies. Moreover, comparative genomics across strains from different regions informs the prediction of emerging pathotypes, enabling preemptive management measures in rice-growing areas.

Advances in gene-editing technologies such as CRISPR-Cas can be applied to either the host or pathogen to disrupt effector binding sites or modulate gene expression. These approaches, informed by high-resolution genomic data, hold promise for reducing the impact of bacterial blight while preserving yield and quality.

In summary, while the core genome of *Xoo* remains conserved across countries, strain-specific differences in genome size, IS element load, TAL effector repertoires, and regulatory gene alleles drive regional differences in virulence and host adaptation. This comparative view highlights how genome plasticity underpins the emergence of novel pathotypes in rice-growing regions of Asia.

11 Disease Cycle and Epidemiology

11.1 PERSISTENCE, TRANSMISSION, AND EPIDEMIOLOGY OF *XOO*

Xanthomonas oryzae pv. *oryzae* (*Xoo*) persists across cropping seasons and landscapes through a combination of specialized ecological strategies and opportunistic exploitation of rice-production practices. Its survival between crop cycles, coupled with efficient dispersal pathways, enables the pathogen to establish recurrent epidemics in diverse agroecosystems. The epidemiology of bacterial blight is shaped by the multiplicative interaction of inoculum reservoirs, transmission routes, weather patterns, cropping intensity, hydrological connectivity, and human-mediated dissemination. Figure 11.1 summarizes the disease cycle, illustrating how primary inoculum sources initiate infection, how secondary spread propagates disease within fields, and how environmental conditions regulate epidemic development. A detailed understanding of these processes remains central to designing effective and durable management strategies for bacterial blight.

11.1.1 Primary Reservoirs: Where *Xoo* Survives between Crops

The persistence of *Xoo* during the off-season is a key determinant of epidemic potential in subsequent crops. While bacterial pathogens generally lack long-lived spores, *Xoo* compensates through survival in substrates that remain biologically active or physically protected in typical rice-growing areas. The longevity of viable bacterial cells varies widely depending on microbial competition, moisture, temperature, and shielding from UV exposure; however, under favorable microhabitats, *Xoo* can survive for several months and occasionally longer in association with plant material and soil organic matter.

- **Seeds**: Seeds represent one of the most critical reservoirs. *Xoo* can colonize the seed surface or become lodged inside the glumes and vascular traces during the grain-filling stage, particularly when infection occurs near panicle emergence. Although internal seed infection tends to occur at lower frequencies, even minimal contamination levels become epidemiologically important in densely sown nurseries where microclimates of high relative humidity and prolonged wetness favor rapid bacterial multiplication. Seed infection does not typically result in systemic infection of seedlings but contributes significantly to early-season spread by producing localized inoculum foci that expand as seedlings are transplanted into main fields.
- **Crop residues and stubble:** Crop residues and stubble are among the most persistent and widely distributed sources of inoculum. After harvest,

DOI: 10.1201/9781003714590-11

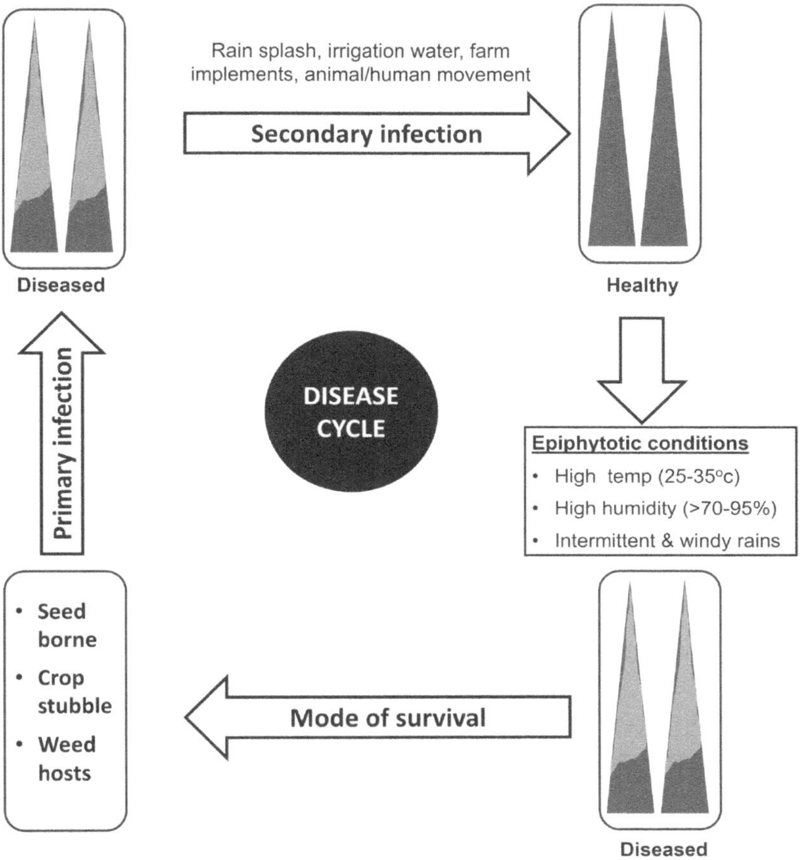

FIGURE 11.1 Disease cycle of bacterial blight of rice.

infected leaves and culms harbor viable bacterial cells within dehydrated lesions, especially near lesion margins where bacterial biomass is highest. Straw left on the soil surface decomposes slowly in many rice systems, particularly in dry-season or rainfed ecologies. Bacteria remain protected in plant tissues where moisture persists or is periodically restored by dew formation. In irrigated or flooded fields, residues soak in standing water, and the bacterial ooze released from macerated tissues contributes to contamination of irrigation water. As residues are transported by tillage operations or moved between fields through water flows, the pathogen can be redistributed across the landscape.

- **Volunteer rice plants:** Volunteer rice plants emerging from shattered grains or ratoon growth act as bridging hosts between cropping seasons. These plants often grow along bunds, canal margins, and neglected field edges where they escape normal crop sanitation. Volunteers that become infected late in the season or harbor latent infections can survive into subsequent seasons, maintaining low but epidemiologically significant inoculum levels. In double- and triple-cropping systems, where fallow periods are short, volunteers become particularly important in maintaining continuity of pathogen populations.

- **Weeds and collateral hosts:** Weeds and collateral hosts are increasingly recognized as long-term reservoirs supporting *Xoo* survival. Numerous grass species, including *Leersia hexandra*, *Echinochloa colona*, *Echinochloa crus-galli*, *Fimbristylis miliacea*, and *Cynodon dactylon*, sustain detectable populations of *Xoo* on their leaf surfaces or within vascular tissues. Although infection on these weeds is often asymptomatic or mild, the plants serve as critical ecological buffers that enable the pathogen to persist in field margins, overflow channels, drainage ditches, and natural wetlands. Because many of these species grow year-round, they maintain *Xoo* populations even in regions with extended dry seasons. Their role in harboring diverse genetic variants further contributes to evolutionary dynamics and race emergence.

Together, seeds, residues, volunteers, and weed hosts create a continuous and interconnected network of inoculum sources across landscapes. Their relative importance varies with cropping system, climate, and management practices, but all contribute to sustaining the pathogen beyond the lifespan of the rice crop.

11.2 PATHWAYS OF SPREAD: HOW *XOO* MOVES WITHIN AND ACROSS FIELDS

Xoo dispersal operates at multiple spatial scales, from short-range within-field transmission to long-distance, regionally significant movement. The pathogen's reliance on water and mechanical forces for dissemination aligns with environmental and agronomic conditions typical of rice-growing ecologies.

- **Rain splash and wind-driven rain:** Rain splash and wind-driven rain constitute primary mechanisms of local spread. When lesions exude bacterial ooze, raindrops dislodge and disperse bacterial aggregates onto adjacent leaves or plants. Short distances, often only a few centimeters, are sufficient for infection as bacteria enter through hydathodes or wounds. Under heavy winds, splash dispersal extends further, creating satellite infection points several meters away. Recurrent rainfall events amplify this effect, transforming small infection foci into expanding epidemic fronts. Once the pathogen establishes within a field, secondary spread rapidly intensifies disease severity.
- **Irrigation and surface water:** Irrigation and surface water provide highly efficient pathways for movement between adjacent fields. In irrigated systems dominated by field-to-field gravity flow, bacterial cells and infected debris are transported through inlet and outlet channels. As water moves across paddies, especially during land preparation or mid-season irrigation pulses, large volumes of inoculum are redistributed. Suspended particles,

including soil colloids and plant debris, offer microhabitats that shield bacteria from desiccation and UV radiation. Contaminated water entering downstream fields serves as a potent driver of inter-field transmission.

- **Human activities, machinery, and farm tools:** Human activities, machinery, and farm tools accelerate the spread in production landscapes with high rates of field traffic. Workers moving between infected and healthy fields may inadvertently transport ooze or contaminated water droplets on footwear or clothing. Transplanters, harvesters, weeders, and rotavators often carry small amounts of infected plant material, which can initiate new infection foci when deposited in clean fields. Poor sanitation of nursery tools, seedling trays, and transport crates further intensifies mechanical dissemination.
- **Seed and seedling:** Seed and seedling trade constitutes one of the most significant long-distance pathways for *Xoo* spread beyond local production zones. In many regions, nurseries supply seedlings to multiple villages or districts, and contaminated seed lots can disseminate the pathogen across broad geographic ranges. Seed movement associated with government distributions, relief operations, or commercial hybrid seed trade has occasionally led to the introduction of new strains and races in previously unaffected regions.
- **Extreme weather events**, including typhoons, cyclones, and severe storms, sometimes contribute to long-distance atmospheric movement of bacterial aerosols or contaminated rain droplets. Although such events occur infrequently, their epidemiological impacts can be severe, triggering sudden regional outbreaks or facilitating intercontinental dispersal of highly virulent strains. High winds accompanying storms may also physically damage rice foliage, creating natural wounds that allow easier bacterial entry and accelerating epidemic development following an event.

These diverse transmission pathways interact to ensure that once established in a region, *Xoo* becomes deeply embedded in the spatial and ecological fabric of rice landscapes.

11.3 EPIDEMIOLOGICAL FACTORS REGULATING DISEASE DEVELOPMENT

The intensity and progression of bacterial blight epidemics depend on the interplay of host susceptibility, pathogen virulence, and environmental conditions. Epidemics flourish when climatic parameters align with pathogen biology, facilitating rapid inoculum multiplication and efficient host colonization.

- **Temperature and humidity** exert strong influences on *Xoo* aggressiveness. Optimal infection typically occurs at temperatures between 30 °C and 35 °C, conditions commonly encountered in tropical and subtropical rice-growing regions. High humidity, particularly above 70%, promotes sustained leaf wetness and increases bacterial survival on leaf surfaces. Night-time humidity surges followed by warm daytime temperatures enhance the likelihood of successful infection events.

- **Leaf wetness and rainfall patterns** regulate the timing and frequency of infection cycles. Dew formation is especially influential in early morning hours when bacterial cells on leaf surfaces hydrate and regain metabolic activity. Intermittent rainfall creates alternating wet and dry periods that promote release of bacterial ooze, splash dispersal, and renewed infection cycles. Paradoxically, continuous heavy rainfall may sometimes reduce infection frequency by washing away ooze, whereas moderate, intermittent rains create the most conducive environment for sustained epidemic progression.
- **Cropping intensity and crop stage** significantly modify epidemiological outcomes. Early infections in nurseries or during the tillering stage lead to extensive systemic colonization, resulting in major yield losses. In high-intensity rice systems with minimal fallow intervals, inoculum carryover is enhanced due to persistent volunteers, stubble, and weed hosts. These systems often experience chronic, overlapping cycles of infection, with epidemics recurring season after season unless extensive seed and residue management interventions are implemented.

The interaction of these environmental and agronomic factors creates a dynamic epidemiological landscape that varies across regions and seasons, influencing the severity and spatial scale of bacterial blight epidemics.

11.4 EPIDEMIOLOGICAL IMPLICATIONS FOR MANAGEMENT

Effective management of bacterial blight requires a comprehensive strategy that integrates field-level practices with landscape-scale interventions targeting inoculum sources and transmission pathways. Because *Xoo* persists in multiple reservoirs and spreads through both natural and human-mediated processes, successful control strategies must address not only symptomatic plants but also hidden inoculum sources in surrounding habitats.

Ensuring seed health through the use of certified, pathogen-free seed is among the most effective means for reducing primary inoculum. Proper seed storage, accompanied by seed sanity testing or hot water treatment where applicable, helps eliminate surface contamination and lowers the risk of nursery-based inoculum buildup.

Nursery hygiene plays a critical role. Separating nurseries from infected fields reduces mechanical spread, while maintaining proper drainage prevents water stagnation that could facilitate bacterial proliferation. Removal of volunteers and feral hosts near nurseries further limits inoculum introduction.

Residue and weed management are essential components of landscape-wide disease suppression. Incorporating residues into soil shortly after harvest accelerates decomposition and lowers bacterial survival. Burning is practiced in some regions but may be discouraged due to environmental concerns; nevertheless, strategic residue removal or composting can substantially reduce inoculum loads. Controlling key weed species along bunds, canals, drains, and irrigation channels disrupts pathogen reservoirs that feed epidemics.

Water and equipment sanitation help interrupt mechanical transmission routes. Ensuring that irrigation water entering a field is free from upstream contamination requires coordinated community action, especially in regions irrigated through shared networks. Cleaning tools and machinery after working in infected fields prevents the movement of plant debris and bacterial ooze into clean fields.

Deployment of resistant cultivars remains the cornerstone of long-term management. Gene pyramiding, rotational use of resistance genes, and integration of broad-spectrum resistance alleles help slow race emergence and ensure durability. Resistance deployment at the landscape level, coordinated among farmers in a region, produces a mosaic of selective environments that reduce the likelihood of dominance by any single virulent race.

Finally, weather-informed surveillance enables rapid responses to conditions conducive to epidemic development. Systems that integrate real-time weather data with disease forecasting models can alert growers to heightened risk periods, enabling timely interventions such as strategic field monitoring, rapid roguing of early infection hotspots, and adjustments to irrigation practices.

Together, these interventions highlight the importance of viewing bacterial blight not merely as a field-level disease but as a landscape-scale epidemiological phenomenon requiring coordinated action across communities, irrigation systems, and cropping cycles.

Xoo persists through a diverse set of ecological reservoirs, including seeds, stubble, volunteer rice, and a wide range of weed hosts. It spreads efficiently within and between fields through splash dispersal, irrigation networks, human activities, seed movement, and occasionally extreme weather events. Warm temperatures, high humidity, and intermittent rainfall create the most favorable epidemiological conditions for disease buildup. Effective management relies on integrated approaches that combine seed health, nursery sanitation, residue and weed control, water and tool hygiene, deployment of resistant cultivars, and weather-based forecasting. Recognizing the landscape-scale nature of inoculum persistence and spread provides the foundation for durable suppression of bacterial blight and reduction of recurrent epidemic risk.

12 Disease Management

12.1 HOST RESISTANCE

Exploiting host plant resistance is an effective strategy for managing bacterial blight (BB) in rice (Jiang *et al.*, 2020; Laha, 2024;). To date, 50 BB resistance (*R*) genes (including 18 recessive) have been identified (Tables 12.1 & 12.2), of which 12 have been cloned and characterized. Combining multiple *R* genes has been shown to be more effective in controlling diverse *Xoo* strains (Ellur *et al.*, 2016a, 2016b; Singh *et al.*, 2023; Sagar *et al.*, 2018, 2020).

The two main genes commonly used in rice breeding for bacterial blight resistance are *xa13* and *Xa21*. Many resistant rice varieties have been developed using these genes (Table 12.3). However, using the same gene or gene combination for a long time can lead to the development of new and more virulent races of *Xoo*, which can break this resistance.

To avoid this problem, different resistance (*R*) genes or their combinations should be used in rotation over time. It is also important to know which *R* genes are present in rice varieties grown in different rice-growing regions. This helps in better planning and distribution of resistant varieties (Wang *et al.*, 2020). In general, identifying new sources of resistance and discovering new resistance genes using molecular tools are essential for developing durable disease-resistant rice varieties.

India has made significant progress in developing Basmati rice varieties resistant to bacterial blight (BB) through both conventional breeding and advanced gene-editing technologies. These efforts are particularly crucial for the north-western Basmati-growing regions, which are heavily affected by this destructive disease. The overall aim is to enhance food security, improve productivity, and ensure sustainable rice cultivation.

12.1.1 Conventional Breeding for Resistance

Several improved Basmati and non-Basmati cultivars have been released using marker-assisted breeding to combat BB and associated stresses:

i. **Pusa 1592:** Pusa 1592, developed through intensive breeding efforts, was officially notified by the Central Release Committee in the Gazette on June 11, 2015, for cultivation in the North-West Basmati growing regions. This variety was developed by incorporating two major bacterial blight (BB) resistance genes, *xa13* and *Xa21*, into the genetic background of the popular Pusa Sugandh 5. Pusa 1592 combines superior grain quality with robust resistance to bacterial blight, making it suitable for sustaining high productivity in disease-prone areas.

DOI: 10.1201/9781003714590-12

TABLE 12.1
Resistance Genes in Rice against Bacterial Blight Races

R Gene	Located in Chromosome	Status of Cloning	Doner	Source Isolate/Races	Dominant/ Recessive	Reference
Xal	4	Yes	Kagyoku, Java 14	Japan	Dominant	Sakaguchi (1967), Yoshimura *et al.* (1998)
Xa2	4	No	Rantai Emas II. Tetep	Vietnam	Dominant	Sakaguchi (1967)
Xa3/ Xa26	11	Yes	Wave Aikoku 3	Japan	Dominant	Ezuka *et al.* (1975), Yoshimura *et al.* (1992), Gao *et al.* (2013)
Xa4	11	Yes	TKM6, IR20, IR22, IR72	India	Dominant	Petpisit *et al.* (1977)
xa5	5	Yes	D2192, 1545–339	Bangladesh	Recessive	Petpisit *et al.* (1977), Blair *et al.* (2003)
Xa6/ xa3	11	No	Malagkit Sungsong Zenith	USA	Dominant	Sidhu and Khush (1978)
Xa7	6	No	DZ78, DV85	Bangladesh	Dominant	Sidhu *et al.* (1978); Luo *et al.*, 2021
xa8	7	No	PI231129	USA	Recessive	Singh *et al.* (2002)
Xa9	11	No	Khaolay Nhay	Laos	Dominant	Singh *et al.* (1983), Ogawa *et al.* (1989)
Xa10	11	Yes	Cas 209	Philippines	Dominant	Yoshimura *et al.* (1983)
Xa11	3	No	RP9–3	Philippines	Dominant	Ogawa and Yamamoto (1986), Goto *et al.* (2009)
Xa12	4	No	Kogyoku, Java 14	Japan	Dominant	Ogawa *et al.* (1978)
xa13	8	Yes	BJ1, Chinsarsh Foro II	India	Recessive	Yoshimura *et al.* (1995), Zhang *et al.* (1996)
Xa14	4	No	TN1	China	Dominant	Taura *et al.* (1987)

(Continued)

TABLE 12.1 *(Continued)*
Resistance Genes in Rice against Bacterial Blight Races

R Gene	Located in Chromosome	Status of Cloning	Doner	Source Isolate/Races	Dominant/ Recessive	Reference
xa15	-	No	M41	Japan	Recessive	Nakai *et al.* (1988)
Xa16	-	No	Tetep	Vietnam	Dominant	Sanchez *et al.* (1999)
Xa17	-	No	Asominori	Japan	Dominant	Ogawa *et al.* (1989)
Xa18	-	No	Toyonishiki, Milyang 23, IR24	Japan	Dominant	Ogawa *et al.* (1986)
xa19	-	No	XM5	Philippines	Recessive	Taura *et al.* (1992)
xa20	-	No	XM6	Philippines	Recessive	Taura *et al.* (1992)
Xa21	11	Yes	*O. jongiuraminata*	Africa Mali	Dominant	Khush *et al.* (1990)
Xa22(t)	11	No	Zhachanglong	China	Dominant	Lin *et al.* (1996)
Xa23	11	Yes	*O. rufipogan*	China	Dominant	Zhang *et al.* (1998)
xa24	2	No	DV86, DVXS, Aus 295	Philippines	Recessive	Wu *et al.* (2008)
xa25(t)	12	Yes	HX-3, Minghui 63	China	Recessive	Lee *et al.* (2003)
xa26(t)	11	No	Minghui 63	China	Recessive	Lee *et al.* (2003)
Xa27(t)	6	Yes	mimta IRGC101141, IRBB27	Philippines	Dominant	Amante-Bordeos *et al.* (1992), Lee *et al.* (2003), Gu *et al.* (2004)
xa28(t)	-	No	Lota Sail	Bangladesh	Recessive	Lee *et al.* (2003)
Xa29(t)	1	No	B5	Portugal, Spain	Dominant	Tan *et al.* (2004)
Xa30(t)	11	No	*O. rufipogan* Y238	India	Dominant	Jin *et al.* (2007)
xa31(t)	4	No	Zhachanglong	China	Recessive	Wang *et al.* (2009)
xa32(t)	11	No	C4064	Australia	Recessive	Zheng *et al.* (2009)

TABLE 12.1 *(Continued)*
Resistance Genes in Rice against Bacterial Blight Races

R Gene	Located in Chromosome	Status of Cloning	Doner	Source Isolate/Races	Dominant/ Recessive	Reference
Xa33	7	No	*O. nivara*	Thailand	Dominant	Kumar *et al.* (2012)
Xa33(t)	6	No	Ba7, *O nivara*	Thailand	Dominant	Korinsak *et al.* (2009)
xa34(t)	1	No	*O brachyantha*	Sri Lanka	Recessive	Ram *et al.* (2010)
Xa35(t)	11	No	*O. minuta*	Philippines	Dominant	Guo *et al.* (2010)
Xa36(t)	11	No	C4059	China	Dominant	Miao *et al.* (2010)
Xa38	4	No	*O. nivara* IRGC81825	India	Dominant	Kaur *et al.* (2005), Cheema *et al.* (2008)
Xa39	11	No	FF329	China, Philippines	Dominant	Zhang *et al.* (2014)
Xa40(t)	11	No	IR65482-7-216-1-2	Korea	Dominant	Kim *et al.* (2015)
xa41(t)	11	Yes	Rice germplasm	-	Recessive	Hutin *et al.* (2015b)
xa42	3	No	XM14, a mutant of IR24	Japan	Recessive	Busungu *et al.* (2016)
Xa43(t)	11	No	P8	Japan	Dominant	Kim and Reinke (2019)
xa44(t)	11	No	IR73571-38-11-3-K3	Japan	Recessive	Kim (2018)
xa45(t)	8	No	*O. glaberrima* IRGC102600B	Philippines	Recessive	Neelam *et al.* (2020)
Xa46(t)	11	No	Mutant H120	Japan	Dominant	Chen *et al.* (2020)
Xa47(t)	11	Yes	G252	China	Dominant	Lu *et al.* (2022), Xing *et al.* (2021)
Xa48 (T)	11	No	*Oryza officinalis*	India	Dominant	Sinha *et al.* (2023)
xa49 (t)	11	No	*O. rufipogan*	India	Recessive	Bhatia *et al.* (2024)
Xa50 (t)	11	No	CAAS-11308X315	China	Dominant	Li *et al.* (2025)

TABLE 12.2
Cloned *R* Genes of Bacterial Blight and Cognate *Xoo Avr* Genes

		Cognate *Avr* gene		
Type of Protein	**Gene**	**Gene**	**Encoding Protein**	**Reference**
NBS-LRR/ NB-ARC-LRR	*Xal*	*PthXol/Tal4/Tal9d*	TAL effector	Yoshimura *et al.* (1998)
NBS-LRR	*Xa47*	*PthXol/Tal4/Tal9d?*	TAL effector	Lu *et al.* (2022)
LRR-RLK	*Xa3/Xa26*	*Avr Xa3*	Unknown	Gao *et al.*,(2013)
LRR-RLK	*Xa21*	*RaxX*	Unknown	Song *et al.* (1995), Wang *et al.* (2006)
WAK	*Xa4*	Not Determined	Unknown	Hu *et al.* (2017)
TM	*Xa10*	*AvrXa10*	TAL effector	Tian *et al.* (2014)
TM	*Xa23*	*AvrXa23*	TAL effector	Wang *et al.* (2015)
SWEET protein	*Xa13 (Os SWEET11)*	*PthXol*	TAL effector	Chu *et al.* (2006)
SWEET protein	*Xa25 (Os SWEET13)*	*PthXo2*	TAL effector	Liu *et al.* (2011)
SWEET protein	*Xa41 (Os SWEET14)*	*AvrXa7/PthXo3/ TalC/Tal5*	TAL effector	Hutin *et al.* (2015b)
Apoplast protein	*Xa27*	*AveXa27*	TAL effector	Lee *et al.* (2003)
TFILA (transcription factor)	*Xa5*	*Avrxa5/PthXo7*	TAL effector	Blair *et al.* (2003)

TABLE 12.3
Exploitation of *R* Genes for BB Resistance in Rice

SN.	*R* Genes	Developed Varieties
1.	*xa13, Xa21*	Improved Pusa Basmati 1 (Pusa 1460), Pusa Basmati 1592, Pusa Basmati 1718, Pusa Basmati 1728, Pusa Basmati 1847, 1886, Punjab Basmati 3
2.	*xa5, xa13, Xa21*	Improved Samba Mahsuri
3.	*Xa4, xa13, Xa21*	PR # 121, 122, 123, and 124
4.	*Xa4, xa5, xa13, Xa21*	NH-56
5.	*Xa4, xa5, Xa7, Xa21*	Joythi
6.	*xa5, xa13, Xa21* and *Xa38*	DRR Dhan 53

ii. **Pusa 1718:** The variety Pusa 1718, registered under PPVFRA Regd. No. REG/2019/139, was notified for release in Punjab, Haryana, and Delhi NCR via Gazette notification on August 29, 2017. It carries the bacterial blight resistance genes *xa13* and *Xa21*, providing durable protection against *Xanthomonas oryzae* pv. *oryzae* (*Xoo*). Pusa 1718-14-2-150 (IET 24565) exhibits excellent agronomic performance along with high grain quality, making it a preferred choice for North-Western Basmati belts.
iii. **Pusa 1728:** Pusa 1728 (IET 24573) was notified for cultivation in North-West Basmati growing areas through the Gazette on November 22, 2016, and later registered under PPVFRA Regd. No. REG/2019/140. The variety incorporates *xa13* and *Xa21*, two key bacterial blight resistance genes, within a high-quality Basmati background. Pusa 1728 demonstrates both strong disease resistance and desirable grain characteristics, addressing the dual needs of productivity and market demand.
iv. **Pusa 1847 (IET 27722):** The variety Pusa 1847-12-62-190-39-7-15 (IET 27722) was officially notified on January 3, 2022, for release in Delhi, Punjab, and Western Uttar Pradesh. Recommended by the Varietal Identification Committee (VIC) and assigned a national identity number by NBPGR, this variety also received NBT certification. Pusa 1847 is a multi-disease resistant Basmati rice combining *xa13* and *Xa21* for bacterial blight resistance with *Pi54* and *Pi2* for blast resistance. It has a seed-to-seed maturity of 115–120 days and an average yield potential of 5.7 t/ha, making it an excellent choice for farmers seeking both high yield and disease resilience.
v. **Pusa 1885 (IET 28807):** Pusa 1885-13-125-20-6 (IET 28807) was notified for release on January 3, 2022, in Delhi, Punjab, and Haryana. It carries *xa13* and *Xa21* for bacterial blight resistance alongside *Pi54* and *Pi2* for blast resistance. This variety demonstrates high adaptability and productivity in the Basmati-growing belts while providing strong protection against two of the most significant rice diseases, thereby contributing to sustainable cultivation.
vi. **Pusa 1886 (IET 28808):** Pusa 1886-13-91-26-9 (IET 28808) was released on January 3, 2022, for Haryana and Uttarakhand. Similar to Pusa 1885, it possesses *xa13* and *Xa21* for bacterial blight resistance, along with *Pi54* and *Pi2* for blast resistance. This variety offers improved disease resistance and grain quality, ensuring stable yields under biotic stress conditions prevalent in these regions.

Beyond Basmati, other rice varieties across India have been enhanced for BB resistance:

i. **Pusa Samba 1853:** Pusa Samba 1853, notified on August 31, 2022, for Andhra Pradesh and Telangana, is a high-quality Basmati rice variety developed with a broader disease resistance spectrum. It carries *xa13* and *Xa21* for bacterial blight resistance and three blast resistance genes—*Pi54*, *Pi1*, and *Pita*. This combination provides enhanced protection against multiple pathogens while maintaining desirable agronomic traits and grain quality, making it suitable for commercial cultivation in Southern Basmati-growing areas.

ii. **Sabour Vibhuti Dhan** (Bihar Agricultural University): A semi-dwarf, multi-gene pyramid carrying *Xa4, xa5, xa13,* and *Xa21*. It shows resilience against BB with only 5–10% damage under stress, withstands flooding, and matures in 135–140 days. It yields 55–60 q/ha, with potential up to 85 q/ha, and is under review for national release.
iii. **Improved Tapaswini** (ICAR-NRRI, Cuttack): A four-gene pyramid (*Xa4, xa5, xa13,* and *Xa21*) that also integrates QTLs for submergence (*Sub1*), salinity (*Saltol*), blast (*Pi2, Pi9*), and gall midge resistance (*Gm1, Gm4*), making it a multi-stress-tolerant line with good quality grains.
iv. **Improved Lalat** (ICAR-NRRI): Carries the same four BB resistance genes in the Lalat background, combining disease resistance with good grain quality and adaptability.
v. **CR Dhan 800** (ICAR-NRRI): A Swarna derivative with *xa5, xa13,* and *Xa21*, moderately to strongly resistant to BB. It is adapted to Odisha, with medium-slender grains and strong yield potential for stress-prone lowlands.
vi. **CR Dhan 803 (Trilochan, IET 26398)**: From the Pooja × Swarna-Sub1 background, it carries BB resistance (gene details not fully publicized), is suited for shallow lowlands, and produces short bold grains with good milling quality. Other NRRI releases (CR Dhan 801, 802, 510, 511, 410, 312, 313) also show moderate BB resistance, combined with submergence tolerance, yield stability, and quality traits.
vii. **Improved Samba Mahsuri** (ICAR-IIRR): Developed via marker-assisted pyramiding, this fine-grain variety incorporates *Xa21*, *xa13*, and *xa5*, making it resistant to multiple virulent *Xoo* strains while retaining excellent cooking quality. It yields 5.5–6.0 t/ha and has a low glycemic index (~50.9).
viii. **DRR Dhan 53 (IET 27294)** (ICAR-IIRR): A four-gene pyramid (*Xa21, xa13, xa5,* and *Xa38*). It is semi-dwarf, non-lodging, matures in 130–135 days, yields ~5.5 t/ha, and is recommended for several BB-prone states across India.

Together, these varieties reflect India's integrated strategy of pyramiding resistance genes with other stress-tolerance traits to develop resilient rice cultivars.

12.1.2 Engineering Resistance through Gene Editing

Traditional breeding based on single-resistance genes often becomes ineffective due to rapid pathogen evolution. Genome editing offers a more reliable approach by directly modifying plant genes with high precision. Technologies such as CRISPR/Cas9 have greatly expanded the possibilities for improving rice health, including enhanced resistance to bacterial blight (Wei *et al.*, 2021). Following are the few gene-edited rice lines that shown promises in improving plant health.

i. **Pusa DST Rice 1** (ICAR-IARI, New Delhi): Based on MTU 1010, this line improves yields by 9.66%–30.4% in saline and alkaline soils, with potential yield increases of up to 20%.

ii. **Kamala (DRR Rice 100)** (ICAR-IIRR, Hyderabad): A CRISPR-edited derivative of Samba Mahsuri (BPT 5204). It matures ~20 days earlier (~130 days), improves grain number per panicle, reduces methane emissions, lowers water and fertilizer use, and has stronger stalks that resist lodging. Grain quality remains similar to Samba Mahsuri.
iii. **Editing *SWEET* Promoters for Broad-Spectrum Resistance**

The virulence of *X. oryzae* pv. *oryzae* (*Xoo*) critically depends on transcription activator-like effectors (TALEs), which are injected into host cells where they bind to effector binding elements (EBEs) in promoters of specific host genes. Upon binding, these TALEs activate transcription of one or more of the three *SWEET* sugar uniporter genes *SWEET11a*, *SWEET13*, or *SWEET14*, thereby facilitating sugar efflux into the apoplast, which *Xoo* exploits for its growth and proliferation within host tissues (Chen *et al.*, 2011, 2012; Streubel *et al.*, 2013; Oliva *et al.*, 2019; Wu *et al.*, 2022; Eom *et al.*, 2019).

Although the rice genome contains 21 *SWEET* paralogs, *Xoo* selectively targets only three *Xa13/Os8N3/SWEET11*, *Xa25/SWEET13*, and *Os11N3/OsSWEET14* across different rice varieties (Blanvillain-Baufumé *et al.*, 2017; Wu *et al.*, 2018). Distinct *Xoo* strains induce expression of one or a combination of these *SWEET* genes to promote pathogenesis. Consequently, deciphering the molecular interactions between *Xoo* TALEs and rice *SWEET* promoters is essential for identifying vulnerabilities in susceptible cultivars (Yuan and Wang, 2013).

Natural resistance to bacterial blight (BB) has evolved through nucleotide polymorphisms in EBEs within *SWEET* promoters, such as *xa13* (*SWEET11a*), *xa25* (*SWEET13*), and *xa41(t)* (*SWEET14*) (Liu *et al.*, 2021; Zhou *et al.*, 2015; Hutin *et al.*, 2015b; Zaka *et al.*, 2018). However, this resistance can be overcome by *Xoo* strains that either recognize alternative EBEs in the same promoter or target another *SWEET* gene (Eom *et al.*, 2019). Marker-assisted pyramiding of *xa13*, *xa25*, and *xa41(t)* could potentially confer broad resistance, but such breeding is complicated by the genetic diversity of donor lines and the presence of multiple EBEs in *SWEET14* that limit complete coverage against all known TALE variants.

To address these challenges, CRISPR/Cas9-mediated multiplex genome editing has emerged as a powerful alternative for precisely modifying multiple EBEs across *SWEET11a*, *SWEET13*, and *SWEET14* promoters (Figure 12.1). This approach effectively prevents TALE binding and induction, thereby conferring broad-spectrum, non–race-specific resistance in diverse rice cultivars (Bezrutczyk *et al.*, 2018; Eom *et al.*, 2019; Ni *et al.*, 2021; Oliva *et al.*, 2019; Rifhani *et al.*, 2023; Schepler-Luu *et al.*, 2023; Ramasamy *et al.*, 2025). Infection assays conducted under controlled growth chamber and greenhouse conditions—standard predictive systems used by breeders—demonstrated that multiplex-edited *Kitaake* exhibited resistance to 96 *Xoo* strains from a global collection (Loo *et al.*, 2024; Oliva *et al.*, 2019).

Building upon these successes, multiplex editing has been deployed in foundational breeding lines (mega-varieties), including *IR64*, widely cultivated in Southeast Asia; *Ciherang-Sub1* (Bina Dhan 11), a submergence-tolerant variety

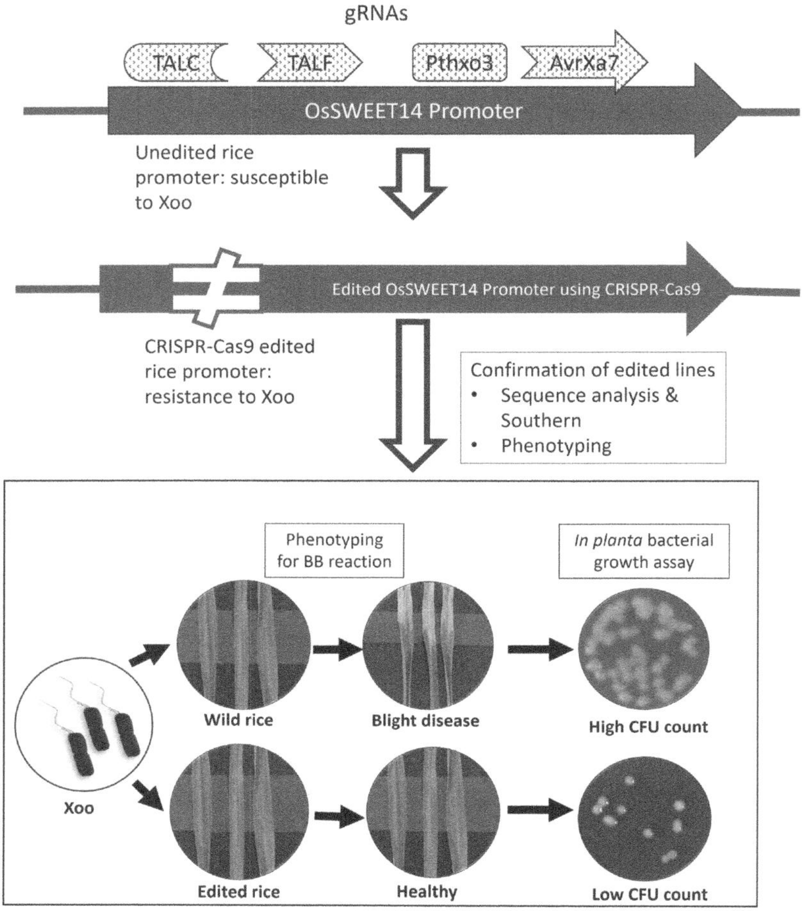

FIGURE 12.1 Editing of rice promoter, specifically EBE region of the *OsSWEET14* promoter, leading to BB resistance.

popular in Indonesia, India, and Bangladesh; and *Komboka*, a leading Kenyan variety (Toledo *et al.*, 2015; Ng'endo *et al.*, 2022). The resulting genome-edited (GE) lines in these elite backgrounds showed robust resistance to representative *Xoo* isolates and exhibited normal growth and agronomic performance under greenhouse and screenhouse conditions (Eom *et al.*, 2019; Oliva *et al.*, 2019; Schepler-Luu *et al.*, 2023).

Comprehensive analyses of these BB-resistant GE lines (*IR64*, *Ciherang-Sub1*, and *Komboka*) were undertaken to facilitate their transfer to small-scale producers in Asia and Africa (Eom *et al.*, 2019; Oliva *et al.*, 2019; Schepler-Luu *et al.*, 2023). A key regulatory requirement for classifying a GE crop as equivalent to conventionally bred lines is the verified absence of foreign DNA or transgene fragments introduced during editing. While absolute proof of absence is technically unattainable due to limitations in detection sensitivity and sequencing biases, careful testing and segregation analyses confirmed effective removal of transgenic elements, making these lines eligible for governmental release. Multi-location field trials in the Philippines and Colombia under non-disease conditions further demonstrated that GE *IR64* and *Ciherang-Sub1* displayed no performance penalties (Loo *et al.*, 2025).

This strategy modulating susceptibility (*S*) gene regulation rather than relying solely on classical resistance (*R*) genes offers a durable, broadly applicable, and environmentally sustainable route to combat bacterial blight in rice. Recent studies confirm that rice lines carrying multiple edited EBEs exhibit significant reductions in BB infection, underscoring the promise of this approach for developing resilient cultivars (Oliva *et al.*, 2019).

In India, the integration of conventional breeding, marker-assisted pyramiding, and CRISPR/Cas9 genome editing is being actively pursued to develop *Basmati* and non-*Basmati* cultivars with enhanced resistance to diverse *Xoo* strains, environmental stresses, and changing climatic conditions. These national initiatives represent a concerted effort to safeguard rice yields, strengthen farmer livelihoods, and ensure long-term food security.

12.2 UNCHARACTERIZED NEW RESISTANCE GENES

OsRPM1 (LOC_Os12g30070.1) encodes a coiled-coil nucleotide-binding leucine-rich repeat (CC-NB-LRR) protein, a structural class of immune receptors that play a central role in effector-triggered immunity (ETI) across diverse plant species. NB-LRR proteins typically act as intracellular sentinels that detect pathogen effectors either through direct recognition or by monitoring effector-induced modifications of host proteins, leading to activation of defense pathways such as reactive oxygen species (ROS) bursts, callose deposition, and in many cases, hypersensitive response (HR)–associated cell death.

Although the function of *OsRPM1* in rice defense had remained unclear for years, recent transcriptome analysis revealed that it is strongly induced during infection by the virulent *X. oryzae* pv. *oryzae* (*Xoo*) race 4 in a type III secretion system (T3SS)-dependent manner (Amrutha Lakshmi *et al.*, 2025). This pattern suggests that *OsRPM1* is activated in response to pathogen-delivered effectors, marking it as a potential participant in ETI. Functional assays using virus-induced gene silencing (VIGS) further established its role: suppression of *OsRPM1* enhanced susceptibility to *Xoo*, with silenced plants showing longer lesions, reduced ROS accumulation, and impaired callose deposition. These phenotypes collectively demonstrate that *OsRPM1* functions as a positive regulator of rice immunity, making it a strong candidate for durable resistance breeding.

Notably, the identification of *OsRPM1* aligns with a growing body of work on newly emerging or poorly characterized *NB-LRR* genes in rice. For example, *Xa33*, originally derived from *Oryza nivara*, encodes a CC-NB-LRR protein conferring resistance to multiple *Xoo* races, although its precise mechanism of effector recognition remains to be elucidated. Similarly, *Xa38*, introgressed from wild rice, has been reported to mediate broad-spectrum bacterial blight resistance, but the downstream signaling pathways are not fully defined. Another case is *Xa46(t)*, a recently mapped resistance locus in rice landraces, which is predicted to encode an NB-LRR protein; however, its effector targets and immune signaling dynamics remain unknown.

These examples underscore the importance of systematically characterizing newly discovered *NB-LRR* genes, since many are identified through genetic mapping or transcriptomics but remain unvalidated at the mechanistic level. Advances in gene editing and effectoromics now provide opportunities to accelerate functional validation of such genes, as shown in the case of *OsRPM1*. By integrating these genes into resistance breeding pipelines, it may be possible to develop cultivars with more durable and broad-spectrum resistance compared to those relying solely on classical TAL effector–targeted *R* genes.

In another study, the transcription factor OsERF101 was functionally verified as an important interacting partner that associates with both *Xa1* and TAL effectors within the nucleus. Functional analyses revealed that altering the expression of *OsERF101*—either through overexpression or gene knockout—enhances Xa1-dependent resistance, indicating a complex regulatory role. Overall, the findings suggest that OsERF101 acts as a key modulator of Xa1-mediated immunity, contributing to the fine-tuning of defense responses in rice against bacterial blight (Yoshihisa *et al.*, 2022).

All these studies highlight a promising direction for rice improvement: moving beyond the traditional R–Avr model toward a deeper understanding of uncharacterized immune receptors, their effector recognition strategies, and their potential to synergize with known quantitative resistance pathways for more sustainable disease control.

12.3 USE OF ANTIBACTERIAL CHEMICALS AND ANTIBIOTICS

Streptocycline is a widely used antibiotic against BB, comprising 90% streptomycin sulfate and 10% tetracycline hydrochloride (Singh *et al.*, 1980). Other effective antibacterial agents include kasugamycin, copper oxychloride, streptomycin sulfate, and a mixture of copper sulfate with lime and water (Table 12.4).

TABLE 12.4
Effectiveness of Chemicals/Antibiotics against BB Pathogen

Name of Chemicals	Effectiveness%	Reference
Copper oxychloride	76.48	Nasir *et al.* (2019)
Streptomycin sulfate	92.23	Nasir *et al.* (2019)
Copper hydroxide (77 % WP)	78.57	Thepbandit *et al.* (2023)

I. **Seed Treatment** Although seed-borne inoculum plays a minor role in disease transmission, infected seeds can introduce the bacterium into new areas. Seed treatment is therefore important. Soaking infected seeds overnight in a 100 ppm streptocycline solution effectively eliminates seed infection.

II. **Root Dip Treatment** Seedlings can be dipped in a 300 ppm streptocycline solution (300 mg/L; 6 g pouch in 20 L water) for 1 hour before transplanting. This treatment helps reduce the incidence of the kresek (yellowing wilt) phase.

III. **Spray Schedule**

1) ***Kresek* phase:** Spray streptocycline at 200 ppm (6 g pouch in 30 L water) if yellowing symptoms appear 7–10 days after transplanting.
2) **Leaf blight phase:** Apply streptocycline at 100 ppm 30–40 days after transplanting. Follow 10–15 days later with a spray of copper oxychloride at 2.5% (2.5 g/L water).
3) **Secondary spread:** Two sprays of Agrimycin-100 at 250 ppm effectively reduce disease intensity and are recommended to control secondary infection.
4) **Field trials:** Experiments at Pantnagar demonstrated that five sprays of Agrimycin-100 and Fytolan (copper oxychloride) in a 50:500 ppm ratio at 12-day intervals significantly reduce disease incidence. Similarly, five sprays of Agrimycin-500 (665 ppm; containing 1.75% streptomycin sulfate, 0.17% terramycin, and 42.4% tribasic copper sulfate) also provide satisfactory disease control.

12.3.1 Antimicrobial Resistance Challenges

Antimicrobial resistance (AMR) arises when bacteria evolve to become less susceptible or entirely resistant to antimicrobial drugs, including antibiotics. This phenomenon occurs through natural selection, where bacterial populations exposed to antimicrobial agents gradually favor the survival of resistant variants. AMR reduces the effectiveness of treatments, making infections harder to manage and increasing the risk of disease spread, severe illness, and death. In agricultural systems, AMR can significantly compromise disease management strategies, leading to crop losses and threatening food security. The emergence of resistance in plant pathogens like *Xoo* could undermine decades of progress in controlling bacterial blight of rice, a major rice disease worldwide.

The use of streptomycin and other bactericides remains an important strategy to complement resistant cultivars and limit the emergence of resistance-breaking *Xoo* strains (Xue, 1973, 2002). In China, bismerthiazol is the most commonly applied bactericide for controlling bacterial blight of rice (Xue, 2002), but frequent and repeated application rapidly selects for bismerthiazol-resistant *Xoo* strains, reducing its long-term efficacy (Wang *et al.*, 1998). Streptomycin, an aminoglycoside antibiotic widely used to treat bacterial infections in humans and animals (Sundin and Bender, 1993), has also been applied in Chinese agriculture to manage bacterial blight of rice (Zhu *et al.*, 1992). Specifically, it has been used in agricultural systems in China for roughly 40 years, with targeted application to prevent rice bacterial blight for approximately 20 years (Xue, 1973). The prolonged use of streptomycin highlights the potential for resistance development, particularly in regions where bactericide application is frequent and environmental conditions favor bacterial proliferation.

In a study conducted in 2007–2008, a total of 534 single-colony *Xoo* isolates were collected from southern China to assess susceptibility to streptomycin. Among these, four isolates (0.75%) from the same county in Yunnan Province exhibited high-level resistance. Remarkably, this resistance could not be attributed to the commonly known genetic determinants for streptomycin resistance in phytopathogenic bacteria, such as the *strA-strB* genes or mutations in the *rpsL* gene (Xu *et al.*, 2010; Sundin, 2000, 2002; Chiou and Jones, 1991, 1995; McManus and Jones, 1994). This finding suggests that alternative mechanisms, possibly involving novel resistance genes or regulatory pathways, may be responsible for the observed resistance, emphasizing the complexity and adaptability of *Xoo* populations.

Resistance integrons, also known as mobile integrons, are key genetic elements that facilitate the acquisition and spread of antibiotic resistance genes among bacterial populations (Stalder *et al.*, 2012; Rowe-Magnus and Mazel, 2002; Zhao *et al.*, 2001). These integrons possess a site-specific recombination system that enables the capture of exogenous open reading frames (ORFs) and their incorporation into the bacterial genome as functional resistance genes (Mazel, 2006; Stokes and Hall, 1989). By serving as natural reservoirs and exchange platforms for resistance determinants, integrons contribute to the rapid evolution of multidrug-resistant bacterial strains in both clinical and agricultural environments. Their presence in phytopathogenic bacteria such as *Xoo* raises concerns about the potential for widespread resistance development, which could compromise the efficacy of key bactericides like streptomycin and bismerthiazol, highlighting the need for integrated disease management strategies that minimize selective pressure and limit the dissemination of resistant strains.

12.4 USE OF BIOCONTROL AGENTS

Biological control has emerged as a sustainable and eco-friendly strategy for managing bacterial blight (BB) of rice caused by *X. oryzae* pv. *oryzae* (*Xoo*). Unlike chemical pesticides, which often contribute to the development of resistant pathogen populations, environmental contamination, and adverse effects on non-target organisms, biocontrol agents (BCAs) offer an effective and ecologically sound alternative. BCAs exploit natural mechanisms such as antibiosis, resource competition, niche exclusion, and activation of host defense pathways to suppress the pathogen and enhance crop health.

Several bacterial and fungal species have been reported to inhibit *Xoo* proliferation, either directly by antagonistic activity or indirectly by modulating host physiology. Direct inhibition is commonly achieved through the secretion of antimicrobial metabolites, siderophores that sequester iron and limit pathogen growth, and hydrolytic or lytic enzymes that degrade cell wall components of the pathogen. Additionally, BCAs occupy ecological niches on the phyllosphere and rhizosphere, thereby competing with *Xoo* for nutrients and spatial dominance (Resti *et al.*, 2020).

Beyond pathogen suppression, many BCAs fall under the broader category of Plant Growth-Promoting Rhizobacteria (PGPR). These organisms not only protect plants from diseases but also improve crop performance through multiple mechanisms, such as nitrogen fixation, phosphate solubilization, phytohormone production, and stress tolerance enhancement. Studies on PGPM typically evaluate parameters such as shoot and root length, total biomass, chlorophyll concentration, photosynthetic efficiency, and the nutritional composition of grains and tissues (Abhilash *et al.*, 2016). Thus, BCAs can simultaneously contribute to disease reduction and yield improvement, offering a dual advantage in sustainable crop production.

The mechanisms underlying BCA activity are diverse and multifaceted. In addition to antibiosis and competition, certain BCAs trigger Induced Systemic Resistance (ISR) or Systemic Acquired Resistance (SAR) in rice, thereby priming the plant immune system for enhanced defense against *Xoo* infection. This systemic activation of host defenses involves pathways regulated by signaling molecules such as salicylic acid, jasmonic acid, and ethylene. The integration of direct antagonism and host-mediated defense priming makes BCAs a promising cornerstone of integrated disease management (IDM) strategies.

The wide range of microorganisms with proven antagonistic activity against *Xoo*, including bacterial genera (*Bacillus, Pseudomonas, Burkholderia, Serratia*) and fungi (*Trichoderma, Aspergillus, Penicillium*), reflects the diversity of natural biocontrol options available. Their reported roles and mechanisms are summarized in Table 12.5 (Mondal *et al.*, 2010), which highlights the major bacterial and fungal species evaluated as biocontrol agents of bacterial blight in rice.

TABLE 12.5
Diverse Mechanisms of BCAs for *Xoo*

Mechanism	BCAs	Principle Compounds	References
Antimicrobial Compounds	*Pseudomonas fluorescens, Bacillus subtilis, Pseudomonas chlororaphis; Bacillus amyloliquefaciens*	• Antibiotics like 2,4diacetylphloroglucinol (DAPG), bacteriocins, and volatile compounds (3-methyl-1-butanol, phenylethyl alcohol, and 2-methyl-1-butanol) that directly inhibit *Xoo*. • Iturins and surfactins (from *Bacillus subtilis*) disrupt bacterial membranes. • Secondary metabolite production of decyl alcohol and 3,5,5-trimethylhexanol, HCN. • Altered surface morphology in the majority of Xanthomonas cells. • Secrete antibiotic compounds difficidin and bacilysin, which distort *Xoo* cell walls.	Yasmin *et al.* (2017),Ryall *et al.* (2009),Velusamy *et al.* (2006),Raaijmakers *et al.* (2002), Ongena and Jacques (2008)Zhang *et al.* (2019), Wu *et al.* (2015a), Shobha *et al.* (2020)

(Continued)

TABLE 12.5 ***(Continued)***
Diverse Mechanisms of BCAs for *Xoo*

Mechanism	BCAs	Principle Compounds	References
Siderophore Production	*Pseudomonas* spp., *Azotobacter* spp.	• Siderophores (phenazine and pyoluteorin) chelate iron from the environment, limiting availability to pathogens.	Haas and Défago (2005), Mishra and Arora (2012)
Nanoparticles	*Trichoderma reesei and Trichoderma harzianum*	ZnO nanoparticles, which have inhibitory effect on *Xoo in vitro*.	
Biofilm Formation	*Brevibacillus laterosporus; Bacillus subtilis; Bacillus amyloliquefaciens*	Biofilm formation inhibits *Xoo* and changes its morphology.	Su *et al.* (2023), Elshakh *et al.* (2016), Xe *et al.* (2018)
Secretion of Lytic Enzymes	*Bacillus* spp., *Streptomyces* spp.	Enzymes like chitinases, glucanases, proteases, and cellulases degrade extracellular polysaccharides and weaken biofilms, impairing colonization and spread of *Xoo*.	Sharma *et al.* (2009), Gopalakrishnan *et al.* (2011)
Competition for Nutrients and Space	*Bacillus* spp., *Pseudomonas* spp., *Trichoderma* spp.	BCAs rapidly colonize niches in the rhizosphere and phyllosphere, limiting *Xoo* access to nutrients and infection sites.	Compant *et al.* (2005), Kloepper *et al.* (2004)
Induction of Systemic Resistance (ISR)	*Pseudomonas fluorescens*, *Bacillus subtilis*	Certain BCAs trigger plant signaling pathways (Jasmonic acid and ethylene), enhancing broad-spectrum resistance against *Xoo* and other pathogens.	Van Loon *et al.* (1998), Pieterse *et al.* (2014)

12.4.1 Production of Antimicrobial Compounds

Many bacterial antagonists produce antibiotics, bacteriocins, and volatile organic compounds (VOCs) that inhibit the growth of *Xoo*. For example, *Pseudomonas fluorescens* secretes phenazine and pyoluteorin, which interfere with pathogen metabolism. Similarly, *Bacillus subtilis* produces iturins and surfactins that compromise bacterial membrane integrity.

I. **Siderophore Production** Iron is an essential nutrient for microbial survival. BCAs such as *Pseudomonas* and *Azotobacter* release siderophores, iron-chelating compounds, that sequester available iron from the rhizosphere, making it inaccessible to *Xoo*. This nutrient competition weakens pathogen growth and reduces infection pressure.
II. **Secretion of Lytic Enzymes** Enzymes such as chitinases, glucanases, proteases, and cellulases degrade structural components of the pathogen's cell wall. Although *Xoo* is not a fungus, certain lytic enzymes can compromise

bacterial extracellular polysaccharides and biofilm formation, disrupting colonization and spread.

III. **Competition for Nutrients and Space** In the rice rhizosphere and phyllosphere, beneficial microbes compete with *Xoo* for carbon sources and colonization sites. Rapid colonizers such as *Bacillus*, *Pseudomonas*, and *Trichoderma* occupy niches on the leaf surface, preventing pathogen attachment and proliferation.

IV. **Induction of Systemic Resistance (ISR)** Some BCAs activate plant defense pathways. For instance, colonization by *Pseudomonas fluorescens* triggers jasmonic acid and ethylene-mediated resistance, enhancing the plant's ability to ward off bacterial invasion. This form of resistance is long-lasting and effective across multiple growth stages.

12.5 ADOPTION OF CULTURAL PRACTICES

Cultural methods constitute one of the most fundamental and farmer-friendly approaches in integrated disease management. These methods involve field management practices that minimize the chances of pathogen survival, multiplication, and subsequent infection in crops. In the case of bacterial blight (BB) of rice, caused by *Xoo*, cultural practices are aimed at eliminating or reducing the primary sources of inoculum that act as reservoirs of the pathogen between cropping cycles. Farmers have traditionally adopted these practices as cost-effective and environmentally sustainable tools to safeguard yield and crop health. According to Shekhar *et al.* (2020), such methods primarily target sources such as infected plant residues, irrigation water, and weed hosts that harbor the pathogen and can become significant contributors to disease outbreaks if ignored.

a) **Management of Infected Plant Material:** Among the various cultural measures, the removal and destruction of infected plant material hold prime importance. Plant debris left in the field after harvest serves as a crucial reservoir of *Xoo*, allowing the bacterium to persist across cropping seasons. If not properly managed, such residues can provide the initial inoculum for the next crop cycle, leading to early and severe infection. Therefore, practices such as burning or deep incorporation of infected stubble and straw can significantly reduce the pathogen load in the field (Shekhar *et al.*, 2020). Moreover, complete drying of fallow fields before sowing disrupts bacterial survival, as *Xoo* is highly dependent on moisture for persistence. By ensuring that no infected material remains on the field surface, the likelihood of pathogen carryover from one season to the next is minimized.

b) **Role of Irrigation Practices and Drainage:** Water management plays a central role in the epidemiology of BB. Irrigation water can serve as a potent carrier of *Xoo*, enabling rapid spread across fields, particularly in regions with interconnected irrigation systems. Hence, designing and maintaining an effective drainage system becomes essential to restrict water-mediated dissemination of the pathogen (Shekhar *et al.*, 2020). Farmers are encouraged to adopt controlled irrigation rather than flooding systems to limit pathogen dispersal. Additionally, avoiding overhead irrigation is crucial, as it prevents the formation of leaf wetness—a condition highly favorable for bacterial multiplication and entry through natural openings or wounds. Instead, irrigation methods such as furrow or drip irrigation, though less commonly used in paddy cultivation, may serve as alternatives in specific farming systems to reduce disease incidence.

c) **Weed Management as a Disease Suppression Strategy:** Weeds play an often-overlooked role in sustaining the *Xoo* population during non-cropping periods. Many grassy weeds and alternate hosts act as symptomless carriers of the bacterium, thereby maintaining a constant inoculum reservoir. Timely and effective weeding, therefore, becomes an indispensable cultural practice to break the disease cycle. Regular weeding not only eliminates alternate hosts but also improves air circulation within the crop canopy, reducing humidity levels that otherwise favor disease development (Shekhar *et al.*, 2020). In modern rice ecosystems where herbicide application is common, integrating chemical weeding with manual removal of potential pathogen hosts provides an additional layer of disease control.

d) **Nutrient Management and Fertilizer Application:** Fertilizer management, particularly with respect to nitrogen, is another factor strongly influencing the susceptibility of rice plants to BB. Excessive application of nitrogen fertilizers promotes lush vegetative growth, resulting in dense canopies that retain moisture and create favorable microclimatic conditions for bacterial proliferation. This not only increases the risk of infection but also supports more rapid spread once the disease sets in. To counter this, balanced nutrient application is recommended, with careful regulation of nitrogen doses and supplementation with phosphorus and potassium for enhanced plant health (Helliwell and Yinong, 2013). In addition, organic amendments that improve soil structure and microbial activity can also suppress pathogen populations indirectly by supporting beneficial antagonistic microbes.

e) **Canopy Management and Aeration:** Canopy structure is closely tied to disease dynamics in rice fields. Dense and unthinned canopies create shaded and humid microenvironments that are highly favorable for *Xoo*. Practices such as maintaining optimum plant spacing, reducing seeding density, and thinning excess tillers help improve air circulation and light penetration within the canopy (Helliwell and Yinong, 2013). These adjustments reduce leaf wetness duration and lower the chances of bacterial entry. Furthermore, soil aeration practices such as periodic harrowing and shallow tillage improve root activity and enhance plant resilience, thereby indirectly reducing disease severity.

f) **Crop Rotation and Fallow Periods:** Crop rotation is another traditional cultural method that plays a role in pathogen suppression. Continuous rice monoculture supports the buildup of *Xoo* inoculum in fields, making subsequent crops increasingly vulnerable. Introducing non-host crops in rotation, such as legumes or oilseeds, interrupts the disease cycle by depriving the pathogen of a suitable host (Helliwell and Yinong, 2013). Additionally, leaving fields fallow and ensuring complete drying during the off-season can further reduce bacterial survival. Although crop rotation is sometimes challenging in rice-dominated ecosystems due to waterlogging requirements and farmer preferences, its integration into disease management plans has shown promising results in reducing BB incidence.

Despite their clear benefits, cultural practices alone often fall short in providing complete control of bacterial blight. With the current rate of re-emergence and adaptation of *Xoo*, these methods are frequently insufficient when applied in isolation (Shekhar *et al.*, 2020). For instance, while proper drainage and weeding reduce pathogen spread, they cannot eliminate inoculum sources completely, especially in regions with high disease pressure and favorable climatic conditions. Similarly, regulated nitrogen use reduces disease severity but cannot fully prevent infection under epidemic scenarios. Hence, cultural methods should be regarded as preventive and suppressive strategies rather than curative solutions.

12.6 INTEGRATION WITH OTHER MANAGEMENT APPROACHES

Given their limitations, cultural methods must be integrated with other disease management strategies, including host plant resistance, biological control agents, and chemical protectants, to achieve sustainable control. The adoption of integrated disease management (IDM) ensures that cultural practices lay the foundation for pathogen suppression, while resistant cultivars and supplementary control measures provide additional layers of protection. Helliwell and Yinong (2013) emphasize that cultural practices enhance the efficiency of other control measures by reducing the initial inoculum load and creating less favorable conditions for pathogen proliferation.

Cultural methods remain indispensable tools for managing bacterial blight of rice, offering farmers a cost-effective, eco-friendly, and easily adoptable set of practices. By addressing primary inoculum sources such as infected residues, contaminated irrigation water, and weed hosts, these methods directly disrupt the disease cycle. Nutrient regulation, canopy management, and crop rotation further contribute to pathogen suppression. However, given the dynamic nature and resilience of *Xoo* populations, cultural practices alone are inadequate in controlling disease occurrence and spread. Their true strength lies in integration with other management strategies under an IDM framework. As noted by Shekhar *et al.* (2020) and Helliwell and Yinong (2013), the effective implementation of cultural practices, when combined with resistant varieties and modern control techniques, offers a promising pathway toward sustainable rice production and long-term disease management.

13 Summary and Way Forward

Rice (*Oryza sativa* L.) is one of the world's most essential cereal crops, forming a primary food source for billions of people globally. With the global population projected to reach nearly 10 billion by 2050, there is an urgent need to increase rice production by approximately 40% to ensure food security (Ray *et al.*, 2013). Among the various biotic constraints that threaten rice production, bacterial blight (BB), caused by *Xanthomonas oryzae* pv. *oryzae* (*Xoo*), stands out as one of the most devastating. This disease has the potential to cause severe yield losses worldwide. In India alone, total crop losses due to BB can range from 50% to complete crop failure, with BB-specific losses reported between 30% and 60%, depending on disease severity (Mondal *et al.*, 2022b; Dubey *et al.*, 2021).

Rice is susceptible to *Xoo* infection at all stages of growth, and the disease manifests in two distinct phases: the *kreshek* phase and the leaf blight phase. The *kreshek* phase primarily occurs in tropical regions during the seedling to early tillering stages and is characterized by sudden yellowing and wilting of seedlings. This phase can lead to total crop failure under severe infections. In contrast, the leaf blight phase, which is more commonly observed during the active tillering stage, begins with water-soaked lesions on leaf blades, usually originating from the leaf tip or margins. These lesions gradually expand downward and inward in a wavy pattern, eventually covering the entire leaf blade within a fortnight or a few weeks, giving the field a characteristic blighted appearance.

Efforts to manage BB have heavily relied on the deployment of resistance genes (*Xa* genes) in rice breeding programs. To date, 50 BB resistance genes have been identified (Ezuka *et al.*, 1975; Petpisit *et al.*, 1977; Khush *et al.*, 1990; Amante-Bordeos *et al.*, 1992; Lin *et al.*, 1996; Blair and McCouch, 1997; Zhang *et al.*, 1996, 1998, 2014; Lee *et al.*, 2003; Gu *et al.*, 2004; Korinsak *et al.*, 2009; Gao *et al.*, 2001, 2013; Chen *et al.*, 2002, 2011, 2020; Blair *et al.*, 2003; Sun *et al.*, 2004; Kaur *et al.*, 2005; Jin *et al.*, 2007; Cheema *et al.*, 2008; Goto *et al.*, 2009; Zheng *et al.*, 2009; Guo *et al.*, 2010; Miao *et al.*, 2010; Ram *et al.*, 2010; Kumar *et al.*, 2012; Busungu *et al.*, 2016; Kim *et al.*, 2015; Kim, 2018; Kim and Reinke, 2019; Neelam *et al.*, 2020; Li *et al.*, 2025). Despite this, the rapid evolution of *Xoo* strains often overcomes these resistant cultivars within just 2–3 years, highlighting the pathogen's adaptability. Globally, over 30 distinct *Xoo* races have been documented (Nino-Liu *et al.*, 2006), and in India, race 4 from the north-western rice-growing states has been identified as a particularly virulent strain. This race has been extensively characterized for its virulence, genomic features, transcriptomic profiles, and type III secretion system (T3SS) effectors (Mondal *et al.*, 2014, 2020c, 2022a).

DOI: 10.1201/9781003714590-13

Xoo employs two types of T3SS-delivered effector proteins to successfully infect rice plants: transcription activator-like effectors (TALEs) and Xanthomonas outer proteins (Xops) (Alfano and Collmer, 2004). Genome-wide screening of *Xoo* race 4 revealed the presence of 18 TALEs and 21 Xop effectors (Mondal *et al.*, 2014). These effectors modulate the rice plant's immune system, either triggering resistance through effector-triggered immunity (ETI) or promoting susceptibility via effector-triggered susceptibility (ETS). Notably, TALEs target susceptibility (*S*) genes such as the *OsSWEET* family by binding to specific effector-binding elements (EBEs) in their promoter regions. Among the five *OsSWEET* genes in rice, *OsSWEET11*, *OsSWEET13*, and *OsSWEET14* are naturally targeted by *Xoo* TALEs, whereas *OsSWEET12* and *OsSWEET15* can act as *S* genes when artificially induced, although no natural *Xoo* strains are known to target them (Antony *et al.*, 2010; Zhou *et al.*, 2015).

To date, six EBEs in these three *OsSWEET* genes have been identified as targets of specific TALEs: PthXo1 binds *OsSWEET11*, PthXo2 binds *OsSWEET13*, and PthXo3, AvrXa7, TalC, and TalF bind *OsSWEET14*. DNA polymorphisms within these EBEs can prevent TALEs from binding to their target promoters, thereby conferring resistance to BB. Consequently, counteracting resistance breakdown by evolving *Xoo* races can be achieved either through the identification of novel durable resistance genes or by modifying susceptible targets using gene-editing technologies.

In this context, the first naturally occurring *OsSWEET* resistance variant, *xa13*, was identified as a promoter mutation in *OsSWEET11* (Chu *et al.*, 2006). The *xa13* allele has demonstrated natural fitness without negatively affecting yield and is extensively used in rice breeding programs to confer BB resistance (Sakthivel *et al.*, 2017). Additional naturally occurring recessive resistance genes have been identified in *OsSWEET13* (*xa25*) and *OsSWEET14* (*xa41*) (Richter *et al.*, 2014; Zhou *et al.*, 2015).

Cultural methods are fundamental, eco-friendly components of integrated disease management (IDM) for bacterial blight (BB) of rice caused by *Xoo*. They primarily aim to reduce inoculum sources such as infected residues, irrigation water, and weed hosts (Shekhar *et al.*, 2020). Key measures include removal or deep incorporation of infected plant material, field drying, and proper drainage to limit water-mediated spread. Weed management is critical, as alternate hosts can sustain *Xoo* populations between seasons. Balanced nutrient management, especially avoiding excessive nitrogen, reduces canopy density and unfavorable microclimates for the pathogen (Helliwell and Yinong, 2013). Proper spacing, tiller thinning, and canopy aeration further minimize humidity and disease incidence. Crop rotation with non-hosts and fallow periods can break the disease cycle, though less commonly practiced in rice-dominated systems (Helliwell and Yinong, 2013).

While effective in reducing inoculum and disease pressure, cultural practices alone are insufficient under high disease pressure and favorable climatic conditions (Shekhar *et al.*, 2020). Their strength lies in integration with resistant cultivars, biocontrol agents, and chemical protectants under an IDM framework, where they enhance the efficiency of other approaches (Helliwell and Yinong, 2013; Shekhar *et al.*, 2020).

Antibiotics and chemicals remain important tools against bacterial blight (BB). Streptocycline (90% streptomycin sulfate, 10% tetracycline hydrochloride) is widely used (Singh *et al.*, 1980), along with kasugamycin, copper oxychloride, and copper sulfate lime mixtures. Seed soaking in 100 ppm streptocycline eliminates seed-borne inoculum. Root-dip treatment with 300 ppm streptocycline reduces kresek incidence. Foliar sprays at different crop stages, streptocycline (100–200 ppm), copper oxychloride (2.5 g/L), and combinations like Agrimycin-100 or Agrimycin-500, have shown significant reduction of disease severity under field conditions.

Frequent and prolonged use of antibiotics and bactericides promotes antimicrobial resistance (AMR), threatening BB management. Streptomycin and bismerthiazol, widely applied in China, have selected for resistant *Xoo* strains, diminishing their long-term efficacy (Xue, 1973, 2002; Wang *et al.*, 1998; Zhu *et al.*, 1992). Surveys of *Xoo* isolates revealed streptomycin-resistant strains not linked to known resistance genes such as *strA-strB* or *rpsL* mutations (Xu *et al.*, 2010; Sundin, 2000, 2002; Chiou and Jones, 1991, 1995; McManus and Jones, 1994). Mobile integrons further accelerate resistance gene acquisition and dissemination (Stalder *et al.*, 2012; Rowe-Magnus and Mazel, 2002; Zhao *et al.*, 2001; Mazel, 2006; Stokes and Hall, 1989). These findings highlight the urgency for integrated disease management to minimize selective pressure and safeguard bactericide efficacy.

Biological control offers a sustainable alternative to chemicals for BB management. Biocontrol agents (BCAs), including *Bacillus*, *Pseudomonas*, *Burkholderia*, *Serratia*, and fungi like *Trichoderma* and *Penicillium*, suppress *Xoo* via antibiosis, siderophore-mediated iron competition, lytic enzyme secretion, niche exclusion, and induction of systemic resistance (Resti *et al.*, 2020; Mondal *et al.*, 2010). Many BCAs also act as Plant Growth-Promoting Rhizobacteria (PGPR), enhancing yield and stress tolerance (Abhilash *et al.*, 2016). Mechanisms include antimicrobial metabolite production, siderophore release, biofilm disruption, and activation of jasmonic acid/ethylene defense pathways. Their integration into IDM provides dual benefits of disease suppression and crop productivity improvement.

This book aims to provide a comprehensive and detailed exploration of bacterial blight of rice, encompassing phenotypic, physiological, genomic, and molecular insights. In particular, it highlights the status of *Xoo* races and presents novel strategies for managing BB based on current knowledge. By focusing on the intricate rice–*Xoo* interactions, this work seeks to uncover new potential targets for disease management and inspire future research to combat this devastating pathogen. With this objective, the author has endeavored to compile a thorough resource that will serve as both a practical guide and a source of inspiration for scientists, breeders, and students working toward sustainable rice production and effective BB management.

Selected References

Abhilash PC, Dubey RK, Tripathi V, Gupta VK and Singh HB 2016. Plant growth-promoting microorganisms for environmental sustainability. *Trends Biotechnol.* 34: 847–850. https://doi.org/10.1016/j.tibtech.2016.05.005.

Akeda Y and Galán JE 2005. Chaperone release and unfolding of substrates in type III secretion. *Nature* 437(7060): 911–915.

Akimoto-Tomiyama C, Furutani A, Tsuge S, Washington EJ, Nishizawa Y, Minami E and Ochiai H 2012. XopR, a type III effector secreted by *Xanthomonas oryzae* pv. *oryzae*, suppresses microbe-associated molecular pattern-triggered immunity in *Arabidopsis thaliana. Molecular Plant Microbe Interactions* 25(4): 505–514.

Alfano JR and Collmer A 2004. Type III secretion system effector proteins: double agents in bacterial disease and plant defense. *Annual Review of Phytopathology* 42: 385–414.

Amante-Bordeos A, Sitch LA, Nelson R, Dalmacio RD, Oliva NP, Aswidinnoor H and Leung H 1992. Transfer of bacterial blight and blast resistance from the tetraploid wild rice *Oryza minuta* to cultivated rice *Oryza sativa. Theoretical and Applied Genetics* 84(3–4): 345–354.

Amrutha Lakshmi M, Kulshreshtha A, Mondal KK, Dasgupta I, Tyagi A, Kumar S, Kalaivanan NS, Mrutyunjaya S, Sreenayana B, Rashmi ER, Ghoshal T, Jagram N, Challa GK and Mani C 2025. Functional validation of OsRPM1 as a positive regulator of bacterial blight resistance in rice via virus--induced gene silencing-. *Folia Microbiologica.* https://doi.org/10.1007/s12223-025-01280-6.

Amrutha Lakshmi M, Mondal KK and Kalaivanan NS 2023. T3SS mediated transcriptional reprogramming of rice by the virulent Indian race 4 of *Xanthomonas oryzae* pv. *oryzae. Oryza* 60: 270–277. https://doi.org/10.35709/ory.2023.60.2.4.

Antony G, Zhou J, Huang S, Li T, Liu B, White F and Yang B 2010. Rice *xa13* recessive resistance to bacterial blight is defeated by induction of the disease susceptibility gene *Os-11N3. The Plant Cell* 22: 3864–3876.

Azizi MMF and Lau HY 2022. Advanced diagnostic approaches developed for the global menace of rice diseases: a review. *Canadian Journal of Plant Pathology* 44(5): 627–651. https://doi.org/10.1080/07060661.2022.2053588

Bai J, Choi SH, Ponciano G, Leung H and Leach JE 2000. *Xanthomonas oryzae* pv. oryzae avirulence genes contribute differently and specifically to pathogen aggressiveness. *Molecular Plant–Microbe Interactions* 13: 1322–1329.

Bezrutczyk M, Yang J, Eom JS, Prior M, Sosso D, Hartwig T, Szurek B, Oliva R, Vera-Cruz C, White FF, Yang B and Frommer WB 2018. Sugar flux and signaling in plant–microbe interactions. *Plant Journal* 93: 675–685.

Bhagya SD, Shanthinie A, Vignesh P, Varanavasiappan S, Kumar KK, Arul L, Kokiladevi E, Manonmani S, Saranya N and Sudhakar D 2023. Targeted editing of OsSWEET11 promoter for imparting bacterial leaf blight resistance in rice. *Electronic Journal of Plant Breeding* 14: 938–947. https://doi.org/10.37992/2023.1403.106.

Bhatia SK, Vikal Y, Kaur P, Dhillon GS, Kaur G, Kumari N, Malik P, Lore JS, Khanna R and Singh K 2024. Introgression and mapping of a novel bacterial blight resistance gene *xa49(t)* from *Oryza rufipogon* acc. CR100098A into *O. sativa. Phytopathology* 114(11): 2412–2420. https://doi.org/10.1094/PHYTO-02-24-0061-R.

Blair MW, Garris AJ, Iyer AS, Chapman B, Kresovich S and McCouch SR 2003. High resolution genetic mapping and candidate gene identification at the *xa5* locus for bacterial blight resistance in rice (*Oryza sativa* L.). *Theoretical and Applied Genetics* 107(1): 62–73.

Blair MW and McCouch SR 1997. Microsatellite and sequence-tagged site markers diagnostic for the rice bacterial leaf blight resistance gene *xa-5*. *Theoretical and Applied Genetics* 95: 174–184.

Blanvillain-Baufumé S, Reschke M, Solé M, Auguy F, Doucoure H, Szurek B and Koebnik R 2017. Targeted promoter editing for rice resistance to *Xanthomonas oryzae* pv. *oryzae* reveals differential activities for SWEET 14-inducing TAL effectors. *Plant Biotechnology Journal* 15: 306–317.

Boch J and Bonas U 2010. Xanthomonas AvrBs3 family-type III effectors: discovery and function. *Annu Rev Phytopathol.* 48: 419–436. https://doi.org/10.1146/annurev-phyto-073009-114447.

Boch J, Scholze H, Schornack S, Landgraf A, Hahn S, Kay S, Lahaye T, Nickstadt A and Bonas U 2009. Breaking the code of DNA binding specificity of TAL-type III effectors. *Science* 326(5959): 1509–1512.

Bogdanove AJ, Koebnik R, Lu H, Feng J and Verdier V 2011. Genome sequencing reveals insights into the evolution and function of *Xanthomonas oryzae* pv. oryzae. *Rice* 4: 54–65.

Bogdanove AJ, Schornack S and Lahaye T 2010. TAL effectors: Finding plant genes for disease and defense. *Current Opinion in Plant Biology* 13: 394–401.

Bokura U 1911. Bacterial leaf blight of rice [in Japanese]. *Teikoku Nokaiho* 2: 62–66.

Buddenhagen IW, Vuong HH and Ba DD 1979. Bacterial leaf blight found in Africa. *International Rice Research Newsletter* 4(1): 11.

Busungu C, Taura S, Sakagami JI and Ichitani K 2016. Identification and linkage analysis of a new rice bacterial blight resistance gene from XM14, a mutant line from IR24. *Breeding Science* 66(4): 636–645.

Büttner D 2012. Behind the lines—actions of bacterial type III effector proteins in plant cells. *FEMS Microbiology Reviews* 40(6): 894–937.

Büttner D and Bonas U 2010. Regulation and secretion of Xanthomonas virulence factors. *FEMS Microbiology Reviews* 34(2): 107–133.

Büttner D and He SY 2009. Type III protein secretion in plant pathogenic bacteria. *Plant Physiology* 150(4): 1656–1664. https://doi.org/10.1104/pp.109.139089

Büttner D, Lorenz C, Weber E and Bonas U 2006. Targeting of two effector protein classes to the type III secretion system by a HpaC- and HpaB-dependent protein complex from *Xanthomonas campestris* pv. *vesicatoria*. *Molecular Microbiology* 59(2): 513–527.

Büttner D, Nennstiel D, Klüsener B and Bonas U 2002. Functional analysis of HrpF, a putative type III translocon protein from *Xanthomonas campestris* pv. *vesicatoria*. *Journal of Bacteriology* 184(9): 2389–2398.

Büttner D, Noël LD and Bonas U 2006. HrcU of *Xanthomonas campestris* pv. *vesicatoria* is an essential T3SS component switching substrate specificity. *Molecular Microbiology* 59: 31–45.

Cheema KK, Grewal NK, Vikal Y, Sharma R, Lore JS, Das A, Bhatia D, Mahajan R, Gupta V, Bharaj TS and Singh K 2008. A novel bacterial blight resistance gene from *Oryza nivara* mapped to 38 kb region on chromosome 4L and transferred to *Oryza sativa* L. *Genetics Research* 90(5): 397–407.

Chen H, Wang S and Zhang Q 2002. New gene for bacterial blight resistance in rice located on chromosome 12 identified from Minghui 63, an elite restorer line. *Phytopathology* 92(7): 750–754.

Chen LQ, Hou BH, Lalonde S, et al 2010. Sugar transporters for intercellular exchange and nutrition of pathogens. *Nature* 468: 527–532.

Chen LQ, Qu XQ, Hou BH, Sosso D, Osorio S, Fernie AR and Frommer WB 2012. Sucrose efflux mediated by SWEET proteins as a key step for phloem transport. *Science* 335: 207–211.

Chen P, Wang J, Liu Q, Liu J, Mo Q, Sun B, Mao X, Jiang L, Zhang J, Lv S, Yu H, Chen W, Liu W and Li C 2024. Transcriptome and metabolome analysis of rice cultivar CBB23 after inoculation by *Xanthomonas oryzae* pv. oryzae strains AH28 and PXO99A. *Plants* 13: 1411. https://doi.org/10.3390/plants13101411.

Chen S, Liu X, Zeng L, Ouyang D, Yang J and Zhu X 2011. Genetic analysis and molecular mapping of a novel recessive gene xa34 (t) for resistance against *Xanthomonas oryzae* pv. *oryzae*. *Theoretical and Applied Genetics* 122: 1331–1338.

Chen S, Wang C, Yang J, Chen B, Wang W, Su J, Feng A, Zeng L and Zhu X 2020. Identification of the novel bacterial blight resistance gene *Xa46 (t)* by mapping and expression analysis of the rice mutant H120. *Scientific Reports* 10: 12642.

Chen X, Liu P, Mei L, He X, Chen L, Liu H, Shen S, Ji Z, Zheng X, Zhang Y, Gao Z, Zeng D, Qian Q and Ma B 2021. *Xa7*, a new executor *R* gene that confers durable and broad-spectrum resistance to bacterial blight disease in rice. *Plant Commun.* 2(3): 100143. https://doi.org/10.1016/j.xplc.2021.100143.

Cheong H, Kim CY, Jeon JS, Lee BM, Sun-Moon J and Hwang I 2013. *Xanthomonas oryzae* pv. *oryzae* type III effector XopN targets OsVOZ2 and a putative thiamine synthase as a virulence factor in rice. *PLoS One* 8(9): e73346.

Chiou CS and Jones AL 1991. The analysis of plasmid-mediated streptomycin resistance in *Erwinia amylovora*. *Phytopathology* 81: 710–714.

Chiou CS and Jones AL 1995. Molecular analysis of high-level streptomycin resistance in *Erwinia amylovora*. *Phytopathology* 85: 324–328.

Cho H, Song ES, Lee YH and Park DS 2008. Comparative analysis of non-TAL type III effectors in Korean *Xanthomonas oryzae* pv. oryzae strains. *Plant Pathology Journal* 24: 1–8.

Chu Z, Fu B, Yang H, Xu C, Li Z, Sanchez A, Park YJ, Bennetzen, JL, Zhang Q and Wang S 2006. Targeting *xa13*, a recessive gene for bacterial blight resistance in rice. *Theoretical and Applied Genetics* 112(3): 455–461.

Compant S, Duffy B, Nowak J, Clément C and Barka EA 2005. Use of plant growth-promoting bacteria for biocontrol of plant diseases: principles, mechanisms of action, and future prospects. *Applied and Environmental Microbiology* 71(9): 4951–4959. https://doi.org/10.1128/AEM.71.9.4951-4959.2005.

Dai X, Wang Y, Yu K, Zhao Y, Xiong L, Wang R and Li S 2023. OsNPR1 enhances rice resistance to *Xanthomonas oryzae* pv. *oryzae* by upregulating rice defense genes and repressing bacteria virulence genes. *International Journal of Molecular Sciences* 24(10): 8687. https://doi.org/10.3390/ijms24108687

Das AK, Saha P, Chattopadhyay S and Chakraborty M 2019. Understanding pathogenesis and host–pathogen interactions in rice bacterial blight. *Plant Science Today* 6: 373–383.

Datta K, Tu J, Oliva N, Ona I, Velazhahan R, Mew TW, Muthukrishnan S and Datta SK 2002. Enhanced resistance to sheath blight and bacterial blight in transgenic rice expressing a chitinase gene. *Plant Science* 163: 313–321.

Dean P 2011. Functional domains and motifs of bacterial type III effector proteins and their roles in infection. *FEMS Microbiology Review* 35(6): 1100–1125. https://doi.org/10.1111/j.1574-6976.2011.00271.x

Deb S, Gupta MK, Patel HK and Sonti RV 2019. *Xanthomonas oryzae* pv. *oryzae* XopQ protein suppresses rice immune responses through interaction with two 14-3-3 proteins but its phospho-null mutant induces rice immune responses and interacts with another 14-3-3 protein. *Molecular Plant Pathology* 20: 976–989.

Dowson WJ 1939. On the systematic position and generic names of the Gram negative bacterial plant pathogens. *Zentralblatt für Bakteriologie, Parasitenkunde und Infektionskrankheiten* 2(100): 177–193.

Dowson WJ 1943. On the generic names *Pseudomonas, Xanthomonas* and *Bacterium* for certain bacterial plant pathogens. *Transactions of the British Mycological Society* 26: 4–14.

Dubey R, Gunasekaran A, Childe SJ, Fosso Wamba S, Roubaud D and Foropon C 2021. Empirical investigation of data analytics capability and organizational flexibility as complements to supply chain resilience. *International Journal of Production Research* 59(1): 110–128.

Duo Z, Liu X, Chen Q and Li Y 2024. Environmental reservoirs of *Xanthomonas oryzae* pv. oryzae in rice-growing ecosystems. *Plant Pathol.* 73: 1125–1136. https://doi.org/10.1111/ppa.14123.

Ejaz K, Zakria M, Zhang P, Huguet TJ, Arif M, White F and Yasmin S 2025. Comparative genomics using long read sequencing identifies nearly identical TAL effector regions in two *Xanthomonas oryzae* pv. oryzae isolates collected from the basmati rice growing region of Pakistan. *Frontiers in Microbiology* 16: 1560969. https://doi.org/10.3389/fmicb.2025.1560969.

Ellur RK, Khanna A, Gopala Krishnan S, Bhowmick PK, Vinod KK, Nagarajan M, Mondal KK, Singh NK, Prabhu KV and Singh AK 2016a. Marker-aided incorporation of *Xa38*, a novel bacterial blight resistance gene, in PB1121 and comparison of its resistance spectrum with *xa13* + *Xa21*. *Scientific Reports* 6: 29188. https://doi.org/10.1038/srep29188.

Ellur RK, Khanna A, Yadav A, Pathania S, Rajashekara H, Singh VK, Gopalakrishnan S, Bhowmick PK, Nagarajan M, Vinod KK, Prakash G, Mondal KK, Singh NK and Singh AK 2016b. Improvement of Basmati rice varieties for resistance to blast and bacterial blight diseases using marker assisted backcross breeding. *Plant Science* 242: 330–341. https://doi.10.1016/j.plantsci.2015.08.020.

Elshakh ASA, Anjum SI, Qiu W, Almoneafy AA, Li W, Yang Z, Cui ZQ, Li B, Sun GC and Xie GL 2016. Controlling and defence-related mechanisms of *Bacillus* strains against bacterial leaf blight of rice. *J Phytopathol.* 164: 534–546. https://doi.org/10.1111/jph.12479.

Eom JS, Luo D, Atienza-Grande G, Yang J, Ji C, Thi-Luu V, Huguet-Tapia JC, Char SN, Liu B, Nguyen H, Schmidt SM, Szurek B, Vera Cruz C, White FF, Oliva R, Yang B and Frommer WB 2019. Diagnostic kit for rice blight resistance. *Nature Biotechnology* 37: 1372–1379.

Espinosa A and Alfano JR 2004. Disabling surveillance: bacterial type III secretion system effectors that suppress innate immunity. *Cell Microbiology* 6: 1027–1040.

Ezuka A, Horino O, Toriyama K, Shinoda H and Morinaka T 1975. Inheritance of resistance of rice variety Wase Aikoku 3 to *Xanthomonas oryzae*. *Bull Tokai-Kinki Natl Agric Exp Stn.* 28: 124–130.

Furutani A, Takaoka M, Sanada H, Noguchi Y, Oku T, Tsuno K, Ochiai H and Tsuge S 2009. Identification of novel type III secretion effectors in *Xanthomonas oryzae* pv. *oryzae*. *Molecular Plant–Microbe Interactions* 22: 96–106.

Furutani A, Tsuge S, Ohnishi K, Hikichi Y and Oku T 2017. HpaB family chaperones in *Xanthomonas* regulate effector secretion through substrate selection. *Frontiers in Microbiology* 8: 1–12.

Furutani A, Tsuge S, Ohnishi K, Hikichi Y, Oku T, Tsuno K, Inoue Y, Ochiai H, Kaku H and Kubo Y 2006. Evidence for HrpX/HrpG-dependent expression of the type III secretion system and effector genes in *Xanthomonas oryzae* pv. *oryzae* under plant-mimicking conditions. *Molecular Plant-Microbe Interactions* 19: 195–203.

Galan JE and Collmer A 1999. Type III secretion machines: bacterial devices for protein delivery into host cells. *Science* 284: 1322–1328.

Gao DY, Xu ZG, Chen ZY, Sun LH, Sun QM, Lu F, Hu BS, Liu YF and Tang LH 2001. Identification of a new gene for resistance to bacterial blight in a somaclonal mutant HX-3 (indica). *Rice Genet Newslett.* 18: 66.

Gao L, Cao Y, Xia Z, Jiang G, Liu G, Zhang W and Zhai W 2013. Do transgenesis and marker-assisted backcross breeding produce substantially equivalent plants? A comparative study of transgenic and backcross rice carrying bacterial blight resistant gene *Xa21*. *BMC Genomics* 14(1): 738.

Garita-Cambronero J, Sena-Vélez M, Palacio-Bielsa A and Cubero J 2016. Genomic islands and T3SS evolution in *Xanthomonas*. *Environmental Microbiology* 18: 1264–1280.

Gnanamanickam SS, Venkatesan B, Narayanan NN, Vasudevan P and Saravana Kumar K 1999. An overview of bacterial blight disease of rice and strategies for its management. *Current Science* 77: 1435–1443.

Gopalakrishnan S, Vadlamudi S, Bandikinda P, Sathya A, Vijayabharathi R, Rupela O, Kudapa H, Katta K and Varshney RK 2011. Evaluation of *Streptomyces* strains isolated from herbal vermicompost for their plant growth-promotion traits in rice. *Microbiological Research* 166(1): 10–20. https://doi.org/10.1016/j.micres.2010.01.003.

Goto T, Matsumoto T, Furuya N, Tsuchiya K and Yoshimura A 2009. Mapping of bacterial blight resistance gene *Xa11* on rice chromosome 3. *Japan Agricultural Research Quarterly: JARQ* 43(3): 221–225.

Grau J, Wolf A, Reschke M, Bonas U, Posch S and Boch J 2019. Computational predictions provide insights into TAL effector target specificity and susceptibility gene regulation in rice. *PLoS Pathog.* 15: e1007621. https://doi.org/10.1371/journal.ppat.1007621.

Gu K, Tian D, Yang FWUL, Wu L, Sreekala C, Wang D, Wang GL and Yin Z 2004. High-resolution genetic mapping of Xa27 (t), a new bacterial blight resistance gene in rice, *Oryza sativa* L. *Theoretical and Applied Genetics* 108: 800–807.

Guo S, DuanPin Z and XingHua L 2010. Identification and mapping of a novel bacterial blight resistance gene Xa35 (t) originated from Oryza minuta. *Scientia Agricultura Sinica* 43(13): 2611–2618.

Guo Y, Figueiredo F, Jones J and Wang N 2012. Characterization of the Xanthomonas citri subsp. citri type III effector PthA4 and its host targets. *Molecular Plant Pathology* 13: 222–235.

Haas D and Défago G 2005. Biological control of soil-borne pathogens by fluorescent pseudomonads. *Nature Reviews Microbiology* 3(4): 307–319. https://doi.org/10.1038/nrmicro1129.

Hacker J and Kaper JB 2000. Pathogenicity islands and the evolution of microbes. *Annu Rev Microbiol.* 54: 641–679. https://doi.org/10.1146/annurev.micro.54.1.641.

Han X, Yang L, Xia F, et al 2025. Rational design of promoter editing confers multipathogen resistance in rice. *Genome Biology* 26: 371. https://doi.org/10.1186/s13059-025-03841-x.

Haq F, Xu X, Ma W, Shah SMA, Liu L, Zhu B, Zou L and Chen G 2022. A Xanthomonas transcription activator-like effector is trapped in nonhost plants for immunity. *Plant Commun.* 3(1): 100249. https://doi.org/10.1016/j.xplc.2021.100249.

Hartmann N, Schulz S, Lorenz C, Fraas S, Hause G and Büttner D 2012. Characterization of HrpB2 from *Xanthomonas campestris* pv. *vesicatoria* identifies protein regions essential for type III secretion pilus formation. *Microbiology* 158: 1334–1349.

Helliwell EE and Yinong Y 2013. Molecular strategies to improve rice disease resistance. In: Y Yang (Ed), *Rice Protocols. Methods in Molecular Biology*, vol. 956. Humana Press, Totowa, NJ, pp. 285–309. https://doi.org/10.1007/978-1-62703-194-3_21.

Hopkins CM, White FF, Choi SH, Guo A and Leach JE 1992. Identification of a family of avirulence genes from *Xanthomonas oryzae* pv. *oryzae*. *Mol Plant Microbe Interact.* 5(6): 451–459. https://doi.org/10.1094/mpmi-5-451.

Hu H, Huang H, Cheng X, Shu X, White AP, Stavrinides J, Köster W, Zhu G, Zhao Z and Wang Y 2017. A global survey of bacterial type III secretion systems and their effectors. *Environmental Microbiology* 19(10): 3879–3895.

Hutin M, Pérez-Quintero AL, Lopez C and Szurek B 2015a. MorTAL Kombat: the story of defense against TAL effectors through loss-of-susceptibility. *Front. Plant Sci.* 6: 535. https://doi.org/10.3389/fpls.2015.00535b

Hutin M, Sabot F, Ghesquière A, Koebnik R and Szurek B 2015b. A knowledge-based molecular screen uncovers a broad-spectrum *OsSWEET14* resistance allele to bacterial blight from wild rice. *Plant J.* 84(4): 694–703. https://doi.org/:10.1111/tpj.13042. PMID: 26426417.

Ikawa Y, Ohnishi S, Shoji A, Furutani A and Tsuge S 2018. Concomitant regulation by a LacI type transcriptional repressor XylR on genes involved in xylan and xylose metabolism and the Type III secretion system in rice pathogen *Xanthomonas oryzae* pv. oryzae. *Molecular Plant Microbe Interactions* 31(6): 605–613.

Ikawa Y and Tsuge S 2016. The quantitative regulation of the hrp regulator HrpX is involved in sugar source dependent hrp gene expression in *Xanthomonas oryzae* pv. *oryzae*. *FEMS Microbiology Letters* 363(10): fnw071. https://doi.org/10.1093/femsle/fnw071

Ishikawa K, Yamaguchi, K, Sakamoto, K, Yoshimura, S, Inoue, K, Tsuge, S, Kojima C and Kawasaki T 2014. Bacterial effector modulation of host E3 ligase activity suppresses PAMP-triggered immunity in rice. *Nature Communication* 5: 5430.

Ishiyama S 1922. Studies of bacterial leaf blight of rice. *Report of the Imperial Agricultural Station, Nishigahara (Konosu)* 45: 233–261.

Iyer AS and McCouch SR 2004. The rice bacterial blight resistance gene *xa5* encodes a novel form of disease resistance. *Molecular Plant-Microbe Interactions* 17(12): 1348–1354.

Jha G, Rajeshwari R and Sonti RV 2007. Functional interplay between two *Xanthomonas oryzae* pv. *oryzae* secretion systems in modulating virulence on rice. *Molecular Plant Microbe Interactions* 20: 31–40.

Ji C, Ji Z, Liu B, Cheng H, Liu H, Liu S, Yang B and Chen G 2020. *Xa1* allelic *R* genes activate rice blight resistance suppressed by interfering TAL effectors. *Plant Commun.* 1(4): 100087. https://doi.org/10.1016/j.xplc.2020.100087.

Ji C, Ji Z, Liu B, Cheng H and Yang B 2016a. TAL effectors that activate OsSWEET genes in rice bacterial blight: diversity and evolutionary dynamics. *Molecular Plant–Microbe Interactions* 29: 914–925.

Ji H, Li T, Li X, Li J, Yu J, Zhang X and Liu D 2022. XopZ and ORP1C cooperate to regulate the virulence of *Xanthomonas oryzae* pv. *oryzae* on Nipponbare. *Plant Signal Behaviour* 17(1): 2035126.

Ji Z, Ji C, Liu B, Zou L, Chen G and Yang B 2016b. Interfering TAL effectors of *Xanthomonas oryzae* neutralize *R*-gene-mediated plant disease resistance. *Nature Communications* 7: 1–9. https://doi.org/10.1038/ncomms13435.

Ji Z, Wang C and Zhao K 2018. Rice routes of countering *Xanthomonas oryzae*. *International Journal of Molecular Sciences* 19: 3008.

Jiang N, Yan J, Liang Y, Shi Y, He Z, Wu Y, Zeng Q, Liu X and Peng J 2020. Resistance genes and their interactions with bacterial blight/leaf streak pathogens (*Xanthomonas oryzae*) in rice (*Oryza sativa* L.)-an updated review. *Rice* 13(1): 3. https://doi.org/10.1186/s12284-019-0358-y.

Jin XW, Wang CL, Yang Q, Jiang JQ, Fan YL, Liu GC and Zhao KJ 2007. Breeding of near-isogenic line CBB30 and molecular mapping of *Xa30 (t)*, a new resistance gene to bacterial blight in rice. *Sci Agric Sin.* 40: 1094–1100.

Jones JD and Dangl JL 2006. The plant immune system. *Nature* 444: 323–329.

Kalaivanan NS, Ghoshal T, Lakshmi Amutha M, Mondal KK, Kulshreshtha A, Singh Kajol BM, Thakur JK, Supriya P, Seema B and Mani C 2023. Complete genome resource unravels the close relation of an Indian *Xanthomonas oryzae* pv. *oryzae* strain IXOBB0003 with Philippines strain causing bacterial blight of rice. *3 Biotech* 13: 187–192. https://doi.org/10.1007/s13205-023-03596-x.

Kauffman HE, Reddy APK, Hsieh SPY and Merca SD 1973. An improved technique for evaluating resistance of rice varieties to *Xanthomonas oryzae*. *Plant Dis Rep*. 57(6): 537–541.

Kaur R, Grewal N, Das A, Vikal Y, Singh J, Bharaj TS, Sidhu JS and Singh K 2005. Inheritance of bacterial blight resistance in two accessions of wild rice, *Oryza nivara*. *Rice Genet News Lett*. 22: 78–82.

Kay S and Bonas U 2009. How Xanthomonas type III effectors manipulate the host plant. *Current Opinion Microbiology* 12: 37–43.

Khush GS, Bacalangco E and Ogawa T 1990. A new gene for resistance to bacterial blight from *O. longistaminata*. *Rice Genet News Lett.* 7: 121–122.

Kim JG, Li X, Roden JA, Taylor KW, Aakre CD, Su B, Lalonde S, Kirik A, Chen Y, Baranage G, McLane H, Martin GB and Mudgett MB 2009. Xanthomonas T3S effector XopN suppresses PAMP-triggered immunity and interacts with a tomato atypical receptor-like kinase and TFT1. *Plant Cell* 21: 1305–1323.

Kim JG, Taylor KW, Hotson A, Keegan M, Schmelz EA and Mudgett MB 2008. XopD SUMO protease affects host transcription, promotes pathogen growth, and delays symptom development in *Xanthomonas*-infected tomato leaves. *Plant Cell* 20: 1915–1929.

Kim SM 2018. Identification of novel recessive gene *xa44* (t) conferring resistance to bacterial blight races in rice by QTL linkage analysis using an SNP chip. *Theoretical and Applied Genetics* 131: 2733–2743.

Kim SM and Reinke RF 2019. A novel resistance gene for bacterial blight in rice, *Xa43* (t) identified by GWAS, confirmed by QTL mapping using a bi-parental population. *PLoS One* 14: e0211775.

Kim SM, Suh JP, Qin Y, Noh TH, Reinke RF and Jena KK 2015. Identification and fine-mapping of a new resistance gene, *Xa40*, conferring resistance to bacterial blight races in rice (*Oryza sativa* L.). *Theoretical and Applied Genetics* 128: 1933–1943.

Kirti K, Rajpal N, Yadav J and Mondal KK 2022. Multi-resolution analysis and deep neural network architecture-based hybrid feature extraction technique for plant disease identification and severity estimation. *Evolutionary Intelligence* https://doi.org/10.1007/s12065-022-00800-4.

Kloepper JW, Ryu CM and Zhang S 2004. Induced systemic resistance and promotion of plant growth by *Bacillus* spp. *Phytopathology* 94(11): 1259–1266. https://doi.org/10.1094/PHYTO.2004.94.11.1259.

Koebnik R, Krüger A, Thieme F, Urban A and Bonas U 2006. Specific binding of the *Xanthomonas campestris* pv. *vesicatoria* AraC-type regulator HrpX to plant-inducible promoter (PIP) boxes. *Journal of Bacteriology* 188: 7652–7660.

Kong W, Ding L and Xia X 2020. Identification and characterization of genes frequently responsive to *Xanthomonas oryzae* pv. *oryzae* and *Magnaporthe oryzae* infections in rice. *BMC Genom.* 21(1): 21. https://doi.10.1186/s12864-019-6438-y.

Korinsak S, Sriprakhon S, Sirithanya P, Jairin J, Korinsak S, Vanavichit A and Toojinda T 2009. Identification of microsatellite markers (SSR) linked to a new bacterial blight resistance gene *xa33* (t) in rice cultivar 'Ba7'. *Maejo Int J Sci Technol.* 3: 235–247.

Kumar PN, Sujatha K, Laha GS, Rao KS, Mishra B, Viraktamath BC, Hari Y, Reddy CS, Balachandran SM, Ram T and Madhav MS 2012. Identification and fine-mapping of *Xa33*, a novel gene for resistance to *Xanthomonas oryzae* pv. *oryzae*. *Phytopathology* 102: 222–228.

Kumar R and Mondal KK 2013. XopN T3SS effector is required for growth and pathogenicity by *Xanthomonas axonopodis* pv. *punicae*, the causal agent of bacterial blight of pomegranate. *Physiological and Molecular Plant Pathology* 84: 36–43.

Kumar R, Soni M and Mondal KK 2016. XopN-T3SS effector of *Xanthomonas axonopodis* pv. *punicae* localizes to the plasma membrane and modulates ROS accumulation events during blight pathogenesis in pomegranate. *Microbiol Res.* 193: 111–120. https://doi.org/10.1016/j.micres.2016.

Laha GS 2024. Geographical Distribution, pathogen diversity and host plant resistance in bacterial blight of rice. *Journal of Mycopathological Research* 62(2): 219–229.

Lee B, Park Y, Park D, Kang H, Kim J, Song E, Park I, Yoon U, Hahn J, Koo B, Lee G, Kim H, Park H, Yoon K, Kim J, Jung C, Koh N, Seo J and Go S 2005. The genome sequence of *Xanthomonas oryzae* pv. *oryzae* KACC10331, the bacterial blight pathogen of rice. *Nucleic Acid Research* 33: 577–586.

Lee KS, Rasabandith S, Angeles ER and Khush GS 2003. Inheritance of resistance to bacterial blight in 21 cultivars of rice. *Phytopathology* 93: 147–152.

Li M, Sohaib SM, Zhou H, Wang J, Liu Y, Wang C, Wang Z and Ji Z 2025. A new bacterial blight resistance gene *Xa50(t)* in the *Xa4* locus confers resistance against *Xanthomonas oryzae* pv. *oryzae* in rice. *Front Plant Sci.* 16: https://doi.org/10.3389/fpls.2025.1657476.

Li R, Wang S, Sun R, He X, Liu Y and Song C 2018. *Xanthomonas oryzae* pv. *oryzae* type III effector PthXo3JXOV suppresses innate immunity, induces susceptibility and binds to multiple targets in rice. *FEMS Microbiology Letters* 365: fny037.

Li T, Liu B, Spalding MH, Weeks DP and Yang B 2016. High-efficiency TALEN-based gene editing produces disease-resistant rice. *Nat Biotechnol.* 34: 336–338. https://doi.org/10.1038/nbt.3480.

Lin XH, Zhang DP, Xie YF, Gao HP and Zhang Q 1996. Identifying and mapping a new gene for bacterial blight resistance in rice based on RFLP markers. *Phytopathology* 86: 1156–1159.

Liu G, Qi Y and Zhang T 2021. Analysis of off-target mutations in CRISPR-edited rice plants using whole-genome sequencing. *Methods in Molecular Biology* 2238: 145–172.

Liu L, Li Y, Wang Q, Xu X, Yan J, Wang Y, Wang Y, Shah SM Ali, Peng Y, Zhu Z, Xu Z and Chen G 2024. Constructed rice tracers identify the major virulent transcription activator like effectors of the bacterial leaf blight pathogen. Rice 17: 30. https://doi.org/10.1186/s12284-024-00704-0.

Liu Q, Yuan M, Zhou Y, Li X, Xiao J and Wang S 2011. A paralog of the MtN3/saliva family recessively confers race-specific resistance to *Xanthomonas oryzae* in rice. *Plant Cell Environ.* 34: 1958–1969.

Long J, Song C, Yan F, Zhou J, Zhou H and Yang B 2018. Non-TAL effectors from *Xanthomonas oryzae* pv. *oryzae* suppress peptidoglycan-triggered MAPK activation in rice. *Frontiers in Plant Science* 9: 1857. https://doi.org/10.3389/fpls.2018.01857.eCollection2018

Loo EPI, Buchholzer M, Huguet-Tapia José C, Selvaraj M, Schepler-Luu V, Rast-Somssich Madlen I, Szurek B, Frommer WB, Stiebner M, Killing B, Hartwig T, Valdéz Gutierrez Sandra P, Tohme J, Charraviaga P, White FF and Yang B 2025. Removal of transgenes and evaluation of yield penalties in genome edited bacterial blight resistant rice varieties. *Plant Biotechnology Journal* 1–15 https://doi.org/10.1111/pbi.70332.

Loo EPI, Szurek B, Arra Y, Stiebner M, Buchholzer M, Devanna BN, Vera Cruz C and Frommer WB 2024. Closing the information gap between the field and scientific literature for improved disease management-with a focus on rice and bacterial blight. *Molecular Plant- Microbe Interactions* 38: 134–141.

Lu W, Pan L, Zhao H, Jia Y, Wang Y, Yu X and Wang X 2014. Molecular detection of *Xanthomonas oryzae* pv. *oryzae*, *Xanthomonas oryzae* pv. *oryzicola*, and *Burkholderia glumae* in infected rice seeds and leaves. *The Crop Journal* 2(6): 398–406. https://doi.org/10.1016/j.cj.2014.06.005.

Lu Y, Zhong Q, Xiao S, Wang B, Ke X, Zhang Y, Yin F, Zhang D, Jiang C, Liu L, Li J, Yu T, Wang L, Cheng Z and Chen L 2022. A new NLR disease resistance gene *Xa47* confers durable and broad-spectrum resistance to bacterial blight in rice. *Front Plant Sci.* 13: 1037901. https://doi.org/10.3389/fpls.2022.1037901.

Luo D, Huguet-Tapia JC, Raborn RT, White FF, Brendel VP and Yang B 2021. The Xa7 resistance gene guards the rice susceptibility gene SWEET14 against exploitation by the bacterial blight pathogen. *Plant Commun.* 2(3): 100164. https://doi.org/10.1016/j.xplc.2021.100164.

Ma W, Xu X, Cai L, Cao Y, Haq F, Alfano JR, Bo Z, Zou L and Chen G 2020. A *Xanthomonas oryzae* type III effector XopL causes cell death through mediating ferredoxin degradation in *Nicotiana benthamiana*. *Phytopathology Research* 2: 16. https://doi.org/10.1186/s42483-020-00055-w

Mak ANS, Bradley P, Bogdanove AJ and Stoddard BL 2013. TAL effectors: function, structure, engineering and applications. *Current Opinion in Structural Biology* 23: 93–99.

Mak ANS, Bradley P, Cernadas RA, Bogdanove AJ and Stoddard BL 2012. The crystal structure of TAL effector PthXo1 bound to its DNA target. *Science* 335: 716–719.

Mazel D 2006. Integrons: agents of bacterial evolution. *Nat Rev Microbiol.* 4: 608–620.

McManus PS and Jones AL 1994. Epidemiology and genetic analysis of streptomycin resistant *Erwinia amylovora* from Michigan and evaluation of oxytetracycline for control. *Phytopathology* 84: 627–633.

Mew TW 1987. Current status and future prospects of research on bacterial blight of rice. *Annual Review of Phytopathology* 25: 359–382. https://doi.org/10.1146/annurev.py.25.090187.002043.

Mew TW 1992. *Bacterial Blight. Compendium of Rice Diseases.* RK Webster and PS Gunnell (Eds). St. Paul: American Phyto pathological Society, pp. 10–11.

Miao L, Wang C, Zheng C, Che J, Gao Y, Wen Y, Li G and Zhao K 2010. Molecular mapping of a new gene for resistance to rice bacterial blight. *Scientia Agricultura Sinica* 43(15): 3051–3058.

Midha S, Bansal K, Kumar S Girija AM, Mishra D, Brahma K, Laha GS, Sundaram RM, Sonti RV and Patil PB 2017. Population genomic insights into variation and evolution of *Xanthomonas oryzae* pv. *oryzae*. *Scientific Reports* 7: 40694 https://doi.org/10.1038/srep40694.

Mishra J and Arora NK 2012. Evaluation of rhizospheric *Pseudomonas* and *Azotobacter* species for their biocontrol and plant growth promoting activities in wheat. *European Journal of Soil Biology* 50: 85–92. https://doi.org/10.1016/j.ejsobi.2011.12.005.

Mondal KK 2011. *Plant Bacteriology*. New Delhi: Kalyani Publishers, p. 200. ISBN: 978-93-272-1631-8.

Mondal KK 2017. Dialogue between Xop-T3SS effectors of *Xanthomonas oryzae* pv. *oryzae* and rice during bacterial blight pathogenesis. *Indian Phytopathology* 70(3): 269–274. https://doi.10.24838/ip.2017.v70.i3.72486.

Mondal KK, Bhar LM and Mani C 2010. Combined efficacy of *Pseudomonas fluorescens* strain MBPF-01 and nanocopper against bacterial leaf blight in rice. *Indian Phytopathology* 63: 266–268.

Mondal KK and Kalaivanan NS 2023. T3SS-effectors of *Xanthomonas oryzae* pv. *oryzae*: The arsenal to bout rice immunity for bacterial blight development. *Oryza* 60: 53–62 (special Issue). https://doi.org/10.35709/ory.2023.60.0.4.

Mondal KK, Kulshreshtha A, Handique PJ, Borbora D, Rajrana Y, Verma G, Bhattacharya A, Qamar A, Lakshmi A, Reddy KK, Soni M, Ghoshal T, Rashmi ER, Mrutynjaya S, Kalaivanan NS and Mani C 2022a. Rice transcriptome upon infection with *Xanthomonas oryzae* pv. *oryzae* relative to its avirulent T3SS-defective strain exposed modulation of many stress responsive genes. *3 Biotech* 12: 130–146. https://doi.org/10.1007/s13205-022-03193-4.

Mondal KK, Laha GS, Mukherjee AK, Prasad MS, Ramanathan A, Rao GP, Krishnaveni D, Matiur R, Bag MK, Bashyal BM and Prakasam V 2022b. *Rice Diseases Compendium*. New Delhi: Indian Phytopathological Society, p. 104. ISBN: 978-81-953723-5-5.

Mondal KK, Meena BR, Junaid A, Verma G, Mani C, Majumder D, Manju Kumar S and Banik S 2014. Pathotyping and genetic screening of type III effectors in Indian strains of *Xanthomonas oryzae* pv. *oryzae* causing bacterial leaf blight of rice. *Physiological and Molecular Plant Pathology* 86: 98–106. https://doi.10.1016/j.pmpp.2014.03.005.

Mondal KK, Rashmi ER and Ghoshal T 2021. Rice SWEET proteins: the key targets by the bacterial blight pathogen, *Xanthomonas oryzae* pv. *oryzae* to ensure nutrition in the rice apoplast during infection. *Indian Phytopathology* 74: 323–331. https://doi.10.1007/s42360-021-00369-5.

Mondal KK, Soni M, Verma G, Kulshreshtha A, Mrutyunjaya S and Kumar R 2020b. *Xanthomonas axonopodis* pv. *punicae* depends on multiple non-TAL (Xop) T3SS effectors for its coveted growth inside the pomegranate plant through repressing the immune responses during bacterial blight development. *Microbiological Research* 240: 126560.

Mondal KK, Verma G, Kulshreshtha A, Rajrana Y, Mani C, Soni M, Reddy KK, Ghoshal T, Lakshmi A and Kalaivanan NS 2020c. Complete Genome Sequence of Indian Race 4 of *Xanthomonas oryzae* pv. oryzae, the Causal Agent of Bacterial Blight of Rice. *Molecular Plant-Microbe Interactions* 33(4): 573–575. https://doi.10.1094/MPMI-12-19-0335-A.

Mondal KK, Verma G, Manju J and Mani C 2016. Rice pathogen *Xanthomonas oryzae* pv. *oryzae* employs inducible *hrp*-dependent XopF type III effector protein for its growth, pathogenicity and for suppression of PTI response to induce blight disease. *European Journal of Plant Pathology* 144: 311–323. https://doi.10.1007/s10658-015-0768-7.

Moscou MJ and Bogdanove AJ 2009. A simple cipher governs DNA recognition by TAL effectors. *Science* 326: 1501. https://doi.org/10.1126/science.1178817.

Nakai H, Nakamura K, Kuwahara S and Saito M 1988. Genetic studies of an induced rice mutant resistant to multiple races of bacterial leaf blight. *Rice Genetics Newsletter* 5: 101–103.

Nakata, Y 1927. On the bacterial leaf blight of rice plant. *Journal of the Society of Tropical Agriculture* 1: 1–16.

Nasir M, Iqbal B, Hussain M, Mustafa A and Ayub M 2019. Chemical management of bacterial leaf blight disease in rice. *J Agric Res*. 57: 99–103.

Neelam K, Mahajan R, Gupta V, Bhatia D, Gill BK, Komal R, Lore JS, Mangat GS and Singh K 2020. High-resolution genetic mapping of a novel bacterial blight resistance gene *xa-45* (t) identified from *Oryza glaberrima* and transferred to *Oryza sativa*. *Theoretical and Applied Genetics* 133(3): 689–705.

Nganga E, Brottier L, Blasco C, Boulard G, Auguy F, Szurek B and Bigirimana J 2025. First report of bacterial leaf blight disease of rice caused by *Xanthomonas oryzae* pv. oryzae in Kenya. *Plant Disease* 109(6): 1373.

Ng'endo M, Kinyua M, Chebet L, Mutiga S, Ndung'u J, Nyongesa O, Njau S, Panchbhai A, Ruth M and Murori R 2022. The importance of market signals in crop varietal development: lessons from komboka rice variety. *CABI Agriculture and Bioscience* 3: 57.

Ni Z, Cao Y, Jin X, Fu Z, Li J, Mo X, He Y, Tang J and Huan S 2021. Engineering resistance to bacterial blight and bacterial leaf streak in rice. *Rice* 14: 38.

Ninc-Liu Do, Ronald PC and Bogdanove AJ 2006. *Xanthomonas oryzae* pathovars: model pathogens of a model crop. *Molecular Plant Pathology* 7: 303–324.

Noël L, Thieme F, Nennstiel D and Bonas U 2001. cDNA-AFLP analysis unravels a genome-wide *hrpG* regulon in the plant pathogen *Xanthomonas campestris* pv. *vesicatoria*. *Molecular Microbiology* 41: 1271–1281.

Ochiai H, Inoue Y, Takeya M, Sasaki A and Kaku H 2005. Genome sequence of *Xanthomonas oryae* pv. *oryzae* suggests contribution of large numbers of effector genes and insertion sequences to its race diversity. *Japan Agricultural Research Quarterly* 39: 275–287.

Ofori AD, Zheng T, Titriku JK, Appiah C, Xiang X, Kandhro AG, Ahmed MI and Zheng A 2025. The role of genetic resistance in rice disease management. *International Journal of Molecular Sciences* 26(3): 956.

Ogawa T, Kaku H and Yamamoto T 1989. Resistance gene of rice cultivar, Asominori to bacterial blight of rice. *Jpn J Breed*. 39(1): 196–197.

Ogawa T, Morinaka T, Fuji K and Kimura T 1978. Inheritance of resistance of rice varieties Kogyoku and Java 14 to bacterial group V of *Xanthomonas oryzae*. *Japanese Journal of Phytopathology* 44(2): 137–141.

Ogawa T, Tabien RE, Busto GA, Khush GS and Mew TW 1986. The relationships between *Xa-3, Xa-4*, and *Xa-4b* for resistances to rice bacterial blight. *Rice Genet Newsletter* 3: 83–84.

Ogawa T and Yamamoto T 1986. Inheritance of resistance to bacterial blight in rice. *Rice Genetics* 471–479.

Oliva R, Ji C, Atienza-Grande G, Huguet-Tapia JC, Perez-Quintero A, Li T, Eom JS, Li C, Nguyen H, Liu B, Auguy F, Sciallano C, Luu VT, Dossa GS, Cunnac S, Schmidt SM, Slamet-Loedin IH, Vera Cruz C, Szurek B, Frommer WB, White FF and Yang B 2019. Broad-spectrum resistance to bacterial blight in rice using genome editing. *Nature Biotechnology* 37: 1344–1350.

Ongena M and Jacques P 2008. *Bacillus* lipopeptides: versatile weapons for plant disease biocontrol. *Trends in Microbiology* 16(3): 115–125. https://doi.org/10.1016/j.tim.2007.12.009.

Petpisit V, Khush GS and Kauffman HE 1977. Inheritance of resistance to bacterial blight in rice 1. *Crop science* 17(4): 551–554.

Pieterse CM, Zamioudis C, Berendsen RL, Weller DM, Van Wees SC and Bakker PA 2014. Induced systemic resistance by beneficial microbes. *Annual Review of Phytopathology* 52: 347–375. https://doi.org/10.1146/annurev-phyto-082712-102340.

Poplawsky AR, Urban SC and Chun W 2000. Biological role of xanthomonadin pigments in *Xanthomonas campestris* pv. *campestris. Applied and Environmental Microbiology* 66(12): 5123–5130

Popov G, Majhi BB and Sessa G 2018. Effector Gene *xopAE* of *Xanthomonas euvesicatoria* 85–10 is part of an operon and encodes an E3 Ubiquitin ligase. *Journal of Bacteriology* 200: e00104-18.

Poulin L, Grygiel P, Magne M, Gagnevin L and Pruvost O 2014. New insights into the evolutionary pathway of *Xanthomonas oryzae* pv. oryzae based on genome sequencing. *Applied and Environmental Microbiology* 80: 595–606.

Pruneda JN, Durkin CH, Geurink PP, Ovaa H, Santhanam B, Holden DW and Komander D 2016. The molecular basis for ubiquitin and ubiquitin-like specificities in bacterial effector proteases. *Molecular Cell* 63: 261–276.

Qi Y, Rao Q, Lu C, Gong J and Hou Y 2025. Recent progress in rice–*Xanthomonas oryzae* interactions. *Biology* 14: 471. https://doi.org/10.3390/biology14050471.

Raaijmakers JM, Vlami M and de Souza JT 2002. Antibiotic production by bacterial biocontrol agents. *Antonie van Leeuwenhoek* 81(1): 537–547. https://doi.org/10.1023/A:1020501420831.

Rajarajeswari NVL and Muralidharan K 2006. Assessments of farm yield and district production loss from bacterial leaf blight epidemics in rice. *Crop Protection* 25: 244–252. https://doi.10.1016/j.cropro.2005.04.013.

Ram T, Laha GS, Gautam SK, Deen R, Madhav MS, Brar DS and Viraktamath BC 2010. Identification of new gene introgressed from *Oryza brachyantha* with broad-spectrum resistance to bacterial blight of rice in India. *Rice Genetics Newsletter* 25: 57.

Ramasamy SP, Rathnasamy SA, Manickam S, Chellappan G, Natarajan B, Swaminathan M, Geetha S and Muthurajan R 2025. Genome editing for improving disease resistance in rice (*Oryza sativa* L.)—scope, achievements, limitations and future needs. *Physiological and Molecular Plant Pathology* 138: 102640.

Rashid MM, Ikawa Y and Tsuge S 2016. GamR, the LysR type galactose metabolism regulator, regulates hrp gene expression via transcriptional activation of two key hrp regulators, HrpG and HrpX, in *Xanthomonas oryzae* pv. oryzae. *Applied and Environmental Microbiology* 82(13): 3947–3958.

Rashmi ER, Mondal KK, Thungri G and Amrutha LM 2024. Comparative gene expression of OsSWEETs from various rice genotypes during the development of bacterial blight as induced by the most virulent Race 4 of *Xanthomonas oryzae* pv. *oryzae*. *Australasian Plant Pathology*. https://doi.org/10.1007/s13313-024-00990-8.

Raveloson H, Rabekijana R, Rakotonanahary NM, Szurek B, Muller B, vom Brocke K and Hutin M 2023. First report of bacterial leaf blight disease of rice caused by *Xanthomonas oryzae* pv. *oryzae* in Madagascar. *Plant Disease* 107(8): 2510.

Ray DK, Mueller ND, West PC and Foley JA 2013. Yield trends are insufficient to double global crop production by 2050. *PloS One* 8(6): p.e66428.

Reddy APK, Mackenzie DR, Rouse DI and Rao AV 1979. Relationship of bacterial leaf blight severity to grain yield of rice. *Phytopathology* 69: 967–969. https://www.apsnet.org/publications/phytopathology/backissues/Documents/1979Articles/Phyto69n09_967.

Reitsma J and Schure J 1950. Bacterial leaf blight of rice in Indonesia. *Contributions from the General Agricultural Research Station* (Bogor, Indonesia) (107): 1–52.

Resti Z, Liswarni Y and Martinius M 2020. Endophytic bacterial consortia as biological control of bacterial leaf blight and plant growth promoter of rice (*Oryza sativa* L.). *JAAST* 4: 134–145. https://doi.org/10.32530/jaast.v4i2.146.

Richter A, Streubel J, Blücher C, Szurek B, Reschke M, Grau J and Boch J 2014. A TAL effector repeat architecture for frameshift binding. *Nature Communications* 5(1): 3447.

Rifhani NF, Apriana A, Sisharmini A, Santoso TJ, Trijatmiko KR, Slamet-Loedin IH and Yunus A 2023. Construction of the CRISPR/Cas9 module and genetic transformation of aromatic rice cv. mentik wangi for developing bacterial leaf blight resistance. *Biodiversitas, Journal of Biological Diversity* 24: 3258–3268.

Rossier O, Wengelnik K, Hahn K and Bonas U 2000. The *hrcC* gene encodes the outer membrane secretin of the *Xanthomonas* T3SS. *Journal of Bacteriology* 182: 1205–1214.

Rowe-Magnus DA, Mazel D 2002. The role of integrons in antibiotic resistance gene capture. *Int J Med Microbiol.* 292: 115–125.

Ryall B, Mitchell H, Mossialos D and Williams HD 2009. Cyanogenesis by the entomopathogenic bacterium *Pseudomonas entomophila*. *Lett App Microbiol.* 49: 131–135. https://doi.org/10.1111/j.1472-765X.2009.02632.x.

Ryan RP, An Sq, Allan JH, McCarthy Y and Dow JM 2015. The DSF family of cell–cell signals: an expanding class of bacterial virulence regulators. *PLoS Pathogens* 11: e1004986.

Ryan RP, Vorhölter FJ, Potnis N, Jones JB, Van Sluys MA, Bogdanove AJ and Dow JM 2011. Pathogenomics of *Xanthomonas*: understanding bacterium–plant interactions. *Nature Reviews Microbiology* 9: 344–355.

Sagar V, Dhawan G, Gopala Krishnan S, Vinod KK, Ellur RK, Mondal KK, Rathour R, Prakash G, Nagarajan M, Bhowmick PK, Bollinedi H and Singh AK 2020. Marker assisted introgression of genes governing resistance to bacterial blight and blast diseases into an elite Basmati rice variety, 'Pusa Basmati 1509'. *Euphytica* 216: 16. https://doi.org/10.1007/s10681-019-2549-4.

Sagar V, Krishnan SG, Dwivedi Priyanka, Mondal KK, Prakash G, Nagarajan M and Singh AK 2018. Development of Basmati rice genotypes with resistance to both bacterial blight and blast diseases using marker assisted restricted backcross breeding. *Indian Journal of Genetics and Plant Breeding* 78: 36–47. https://doi.org/10.5958/0975-6906.2018.00005.6.

Sakaguchi S 1967. Linkage studies on the resistance to bacterial leaf blight, *Xanthomonas oryzae* (Uyeda et Ishiyama) Dowson, in rice. *Bulletin of the National Institute of Agricultural Sciences, Series D* 16: 1–18.

Sakthivel K, Gautam RK, Manigundan K, Singh R, Ramalingam J, Laha GS, Kumar A and Velazhahan R 2017. The host background of rice influences the resistance expression of a three genes pyramid (*xa5* + *xa13* + *Xa21*) to bacterial blight (*Xanthomonas oryzae* pv. *oryzae*) pathotypes of Indian mainland and Bay islands. *Plant Breeding* 136(3): 357–364.

Salzberg SL, Sommer DD, Schatz MC, Phillippy AM, Rabinowicz PD, Tsuge S, Furutani A, Ochiai H, Delcher AL, Kelley D, Madupu R, Puiu D, Radune D, Shumway M, Trapnell C, Aparna G, Jha G, Pandey A, Patil PB, Ishihara H, Meyer DF, Szurek B, Verdier V, Koebnik R, Dow JM, Ryan RP, Hirata H, Tsuyumu S, Lee SW, Seo YS, Sriariyanum M, Ronald PC, Sonti RV, Van Sluys MA, Leach JE, White FF and Bogdanove AJ 2008. Genome sequence and rapid evolution of the rice pathogen *Xanthomonas oryzae* pv. oryzae PXO99A. *BMC Genomics* 9: 204. https://doi.org/10.1186/1471-2164-9-204.

Sanchez AC, Brar DS, Huang N, Li ZK and Khush GS 2000. Sequence tagged site marker-assisted selection for three bacterial blight resistance genes in rice. *Crop Science* 40: 792–797.

Sanchez AC, Ilag LL, Yang D, Brar DS, Ausubel F, Khush GS, Yano M, Sasaki T, Li Z and Huang N 1999. Genetic and physical mapping of xa13, a recessive bacterial blight resistance gene in rice. *Theoretical and Applied Genetics* 98: 1022–1028.

Schepler-Luu V, Sciallano C, Stiebner M, Ji C, Boulard G, Diallo A, Auguy F, Char SN, Arra Y, Schenstnyi K, Buchholzer M, Loo EPI, Bilaro AL, Lihepanyama D, Mkuya M, Murori R, Oliva R, Cunnac S, Yang B, Szurek B and Frommer WB 2023. Genome editing of an African elite rice variety confers resistance against endemic and emerging *Xanthomonas oryzae* pv. *oryzae* strains. *eLife* 12:e84864. https://doi.org/10.7554/eLife.84864.

Sharma A, Johri BN, Sharma AK and Glick BR 2009. Plant growth-promoting bacterium *Pseudomonas* sp. strain GRP3 influences iron acquisition in mung bean (*Vigna radiata* L. Wilzeck). *Soil Biology and Biochemistry* 41(5): 1084–1091. https://doi.org/10.1016/j.soilbio.2009.02.006.

Shekhar S, Sinha D and Kumari A 2020. An overview of bacterial leaf blight disease of rice and different strategies for its management. *Int J Curr Microbiol App Sci.* 9: 2250–2265. https://doi.org/10.20546/ijcmas.2020.904.270.

Shobha B, Lakshmeesha TR, Ansari MA, Almatroudi A, Alzohairy MA, Basavaraju S, Alurappa R, Niranjana SR and Chowdappa 2020. Mycosynthesis of ZnO nanoparticles using *Trichoderma* spp. isolated from rhizosphere soils and its synergistic antibacterial effect against *Xanthomonas oryzae* pv. *oryzae*. *J Fungi*. 6: 181. https://doi.org/10.3390/jof6030181.

Sidhu G, Khush G and Mew T 1978. Genetic analysis of bacterial blight resistance in seventy-four cultivars of rice, *Oryza sativa L. Theoretical and Applied Genetics* 53(3): 105–111.

Sidhu GS and Khush GS 1978. Dominance reversal of a bacterial blight resistance gene in some rice cultivars. *Phytopathology* 68: 461–463.

Singer AU, Schulze S, Skarina T, Xu X, Cui H, Eschen-Lippold L, Egler M, Srikumar T, Raught B, Lee J, Scheel D, Savchenko A and Bonas U 2013. A pathogen type III effector with a novel E3 ubiquitin ligase architecture. *PLoS Pathogens* 9: e1003121.

Singh G, Singh N, Ellur RK, Balamurugan AGP, Rathour R, Mondal KK, Bhowmick PK, Gopalakrishnan S, Nagarajan M, Seth R, Vinod KK, Singh Varsha, Bollinedi H and Singh AK 2023. Genetic enhancement for biotic stress resistance in basmati rice through marker-assisted backcross breeding. *International Journal Molecular Science* 24: x. https://doi.org/10.3390/xxxxx.

Singh K, Vikal Y, Singh S, Leung H, Dhaliwal HS and Khush GS 2002. Mapping of bacterial blight resistance gene *xa8* using microsatellite markers. *Rice Genetics Newsletter* 19: 94–97.

Singh N, Mishra SK, Singh VK and Tiwari S 2016. Marker-assisted selection for pyramiding bacterial blight resistance genes in rice. *Rice* 9: 12. https://doi.org/10.1186/s12284-016-0085-4.

Singh RA, Das B, Ahmed KM and Pal V 1980. Chemical control of bacterial leaf blight of rice. *Int J Pest Manag*. 26: 21–25. https://doi.org/10.1080/09670878009414278.

Singh RJ, Khush GS and Mew TW 1983. A new gene for resistance to bacterial blight in rice 1. *Crop Science* 23: 558–560.

Sinha P, Dilip Kumar TK, Solanki M, Gokulan CG, Das, Miriyala A, Gonuguntala R, Elumalai P, Kousik MBVN, Masthani SK, Kumboju C, Yugander A, Laha GS, Chirravuri NN, Patel HK, Ghazi IA, Kim SR, Jena KK, Hanumanth SR, Ricardo Oliva R, Mangrauthia SK and Sundaram RM 2023. Fine mapping and sequence analysis reveal a promising candidate gene encoding a novel NB-ARC domain derived from wild rice (*Oryza officinalis*) that confers bacterial blight resistance. *Front Plant Sci.* 14: 1173063. https://doi.org/10.3389/fpls.2023.1173063.

Song C and Yang B 2010a. Mutagenesis of 18 type III effectors reveals virulence function of XopZ (PXO99) in *Xanthomonas oryzae* pv. *oryzae*. *Molecular Plant Microbe Interactions* 23: 893–902 https://doi.org/10.1094/MPMI-23-7-0893.

Song WY, Wang GL, Chen LL, Kim HS, Pi LY, Holsten T, Gardner J, Wang B, Zhai WX, Zhu LH, Fauquet C and Ronald PC 1995. A receptor kinase–like protein encoded by the rice disease-resistance gene, Xa21. *Science* 270: 1804–1806. https://doi.org/10.1126/science.270.5243.1804.

Stalder T, Barraud O, Casellas M, Dagot C and Ploy MC 2012. Integron involvement in environmental spread of antibiotic resistance. *Front Microbiol.* 3: 119.

Stokes HW and Hall RM 1989. A novel family of potentially mobile DNA elements encoding site-specific gene-integration functions: integrons. *Mol Microbiol.* 3: 1669–1683.

Streubel J, Pesce C, Hutin M, Koebnik R, Boch J and Szurek B 2013. Five phylogenetically close rice *SWEET* genes confer TAL effector-mediated susceptibility to *Xanthomonas oryzae* pv. *oryzae*. *New Phytologist* 200: 808–819.

Su XX, Wan TT, Gao YD, Zhang SH, Chen X, Huang LQ and Wang W 2023. Action mechanism of the potential biocontrol agent *Brevibacillus laterosporus* SN19–1against *Xanthomonas oryzae* pv. *oryzae* causing rice bacterial leaf blight. *Arch Microbiol.* 206: 40. https://doi.org/10.1007/s00203-023-03754-y.

Sun X, Cao Y, Yang Z, Xu C, Li X, Wang S and Zhang Q 2004. *Xa26*, a gene conferring resistance to *Xanthomonas oryzae* pv. *oryzae* in rice, encodes an LRR receptor kinase like protein. *The Plant Journal* 37(4): 517–527.

Sundin GW 2000. Examination of base pair variants of the strA-strB streptomycin resistance genes from bacterial pathogens of humans, animals and plants. *J Antimicrob Chemoth.* 46: 848–849.

Sundin GW 2002. Distinct recent lineages of the strA-strB streptomycin resistance genes in clinical and environmental bacteria. *Curr Microbiol.* 45: 63 69.

Sundin GW and Bender CL 1993. Ecological and genetic analysis of copper and streptomycin resistance in *Pseudomonas syringae* pv. *syringae*. *Appl Environ Microbiol.* 59: 1018–1024.

Swings J, Van den Mooter M, Vauterin L, Hoste B, Gillis M, Mew TW and Kersters K 1990. Reclassification of the causal agents of bacterial blight (*Xanthomonas campestris* pv. *oryzae*) and bacterial leaf streak (*Xanthomonas campestris* pv. *oryzicola*) of rice as pathovars of *Xanthomonas oryzae* (ex-Ishiyama 1922) sp. nov., nom. rev. *International Journal of Systematic and Evolutionary Microbiology* 40(3): 309–311.

Tan GX, Ren X, Weng QM, Shi ZY, Zhu LL and He GC 2004. Mapping of a new resistance gene to bacterial blight in rice line introgressed from *Oryza officinalis*. *Yi chuan xue bao= Acta Genetica Sinica* 31: 724–729.

Tariq R, Ji Z, Wang C, Tang Y, Zou L, Sun H, Chen G and Zhao K 2019. RNA-Seq analysis of gene expression changes triggered by *Xanthomonas oryzae* pv. *oryzae* in a susceptible rice genotype. *Rice* 12: 44.

Taura S, Ogawa T, Tabien RE, Khush GS, Yoshimura A and Omura T 1987. The specific reaction of Taichung Native 1 to Philippine races of bacterial blight and inheritance of resistance to race 5 (PXO 112). *Rice Genetics Newsletter* 4: 101–102.

Taura S, Ogawa T, Yoshimura A and Ikeda R 1992. Identification of a recessive resistance gene to rice bacterial blight of mutant line XM6, *Oryza sativa* L. *Japanese Journal of Breeding* 42: 7–13.

Thakur A, Yadav R, Singh D and Kumar S 2013. Mapping of resistance loci against *Xanthomonas oryzae* pv. oryzae in Indian rice cultivars. *Plant Dis.* 97: 802–810. https://doi.org/10.1094/PDIS-11-12-0999-RE.

Thepbandit W, Papathoti NK, Daddam JR, Hoang NH, Le Thanh T, Saengchan C and Buensanteai K 2023. *In vitro* and *in silico* studies of salicylic acid on systemic induced resistance against bacterial leaf blight disease and enhancement of crop yield. *J Inte Agri.* 22: 170–184. https://doi.org/10.1016/jia.2022.08.112.

Tian D, Wang J, Zeng X, Gu K, Qiu C, Yang X, Zhou Z, Goh M, Luo Y, Murata-Hori M, White FF and Yin Z 2014. The rice TAL effector-dependent resistance protein XA10 triggers cell death and calcium depletion in the endoplasmic reticulum. *Plant Cell* 26(1): 497–515. https://doi.org/10.1105/tpc.113.119255.

Timilsina S, Potnis N, Newberry EA, Liyanapathiranage P, Iruegas-Bocardo F, White FF, Erica M, Goss EM and Jones JB 2020. *Xanthomonas* diversity, virulence and plant–pathogen interactions. *Nature Reviews Microbiology* 18: 415–427.

Tiwari PK, Suman, Sahu S, Sahu PK, Patel RR, Baghel S and Sharma D 2025. Emergence of a new *Xanthomonas oryzae* pv. oryzae pathotype breaking the earlier resistance gene combination of improved Sambha Mahsuri. *Electronic Journal of Plant Breeding* 16: 197–205. https://doi.org/10.37992/2025.1602.025.

Tiwari S, Singh VK, Kumar A and Pandey A 2015. Pyramiding of bacterial blight resistance genes in high yielding rice cultivars in India. *Eur J Plant Pathol.* 141: 605–616. https://doi.org/10.1007/s10658-014-0567-2.

Toledo AMU, Ignacio JCI, Casal C, Gonzaga ZJ, Mendioro MS and Septiningsih EM 2015. Development of improved ciherang-sub1 having tolerance to anaerobic germination conditions. *Plant Breeding and Biotechnology* 3: 77–87.

Tran TT, Pérez-Quintero AL, Wonni I, Carpenter SC, Yu Y, Wang L, Leach JE and Szurek B 2018. Functional analysis of African *Xanthomonas oryzae* pv. *oryzae* TALomes reveals a new susceptibility gene in bacterial leaf blight of rice. *PLoS Pathogens* 14: e1007092.

Tsuge S, Furutani A, Fukunaka R, Oku T and Tsuno K 2005. PIP-box–mediated regulation of T3SS genes in *Xanthomonas oryzae* pv. *oryzae*. *Journal of Bacteriology* 187: 145–156.

Üstün S, Bartetzko V and Börnke F 2013. The *Xanthomonas campestris* type III effector XopJ proteolytically degrades the proteasome subunit RPT6 and inhibits proteasome activity in plants. *The Plant Cell* 25: 3515–3527.

Van Loon LC, Bakker PAHM and Pieterse CMJ 1998. Systemic resistance induced by rhizosphere bacteria. *Annual Review of Phytopathology* 36: 453–483. https://doi.org/10.1146/annurev.phyto.36.1.453.

Velusamy P, Immanuel JE, Gnanamanickam SS and Thomashow L 2006. Biological control of rice bacterial blight by plant-associated bacteria producing 2,4-diacetylphloroglucinol. *Can J Microbiol.* 52: 56–65. https://doi.org/10.1139/w05-106.

Veracruz CM 1985. *Studies on the bacterial blight of rice (Xanthomonas campestris* pv. *oryzae) in the Philippines.* Ph.D. Thesis, University of the Philippines Los Baños, Los Baños, Philippines.

Verma G, Mondal KK, Kulshreshtha A and Sharma M 2019. XopR T3SS--effector of- *Xanthomonas oryzae* pv. oryzae suppresses cell death--mediated plant defense response during bacterial blight development in rice-. *3 Biotech* 9: 272–281. https://doi.org/10.1007/s13205-019-1802-9.

Verma G, Sharma M and Mondal KK 2018. XopR TTSS-effector regulates *in planta* growth, virulence of Indian strain of *Xanthomonas oryzae* pv. *oryzae* via suppressing reactive oxygen species production and cell wall-associated rice immune responses during blight induction. *Functional Plant Biology* 45: 561–574. https://doi.org/10.1071/FP17147.

Vivian A, Murillo J and Jackson RW 2001. The roles of plasmids in phytopathogenic bacteria: mobile arsenals? *Microbiology* 147(4): 763–780. https://doi.org/10.1099/00221287-147-4-763.

Wang C, Wen G, Lin X, Liu X and Zhang D 2009. Identification and fine mapping of the new bacterial blight resistance gene, *Xa31* (t), in rice. *European Journal of Plant Pathology* 123: 235–240.

Wang C, Zhang X, Fan Y, Gao Y, Zhu Q, Zheng C, Qin T, Li Y, Che J, Zhang M and Yang B 2015. XA23 is an executor R protein and confers broad-spectrum disease resistance in rice. *Molecular Plant* 8: 290–302.

Wang S, Liu W, Lu D, Lu Z, Wang X, Xue J and He X 2020. Distribution of bacterial blight resistance genes in the main cultivars and application of Xa23 in rice breeding. *Front Plant Sci.* 11: 555228. https://doi.org/10.3389/fpls.2020.555228.

Wang WX, Gu JT, Xia J and Vichai K 1998. A research on the resistances of *Xanthomonas oryzae* pv. *oryzae* to bactericides under laboratory conditions. *Acta Phytophylacica Sinica* 25: 171–174.

Wang Y, Liang H, Guo R, Chen L, He C and Jiang W 2018. Functional mapping of HrpE pilin domains in *Xanthomonas*. *Molecular Plant Pathology* 19: 2121–2135.

Wang YS, Wang GL, Pi LY, Walker JC and Song WY 2006. Rice XA21 binding protein 3 is a ubiquitin ligase required for full accumulation of XA21 and for *Xa21*-mediated resistance. *Plant Cell* 18: 3635–3646. https://doi.org/10.1105/tpc.106.047958.

Weber E, Koebnik R, Brisset MN, Neyt C, Taha MK and Bonas U 2005. Structural determinants of HrpE polymerization and T3SS function in *Xanthomonas*. *Journal of Molecular Biology* 353: 1251–1265.

Wei Z, Abdelrahman M, Gao Y, Ji Z, Mishra R, Sun H, Sui Y, Wu C, Wang C and Zhao K 2021. Engineering broad-spectrum resistance to bacterial blight by CRISPR-Cas9-mediated precise homology directed repair in rice. *Mol Plant.* 14(8): 1215–1218. https://doi.org/10.1016/j.molp.2021.05.012.

Wengelnik K, van den Ackerveken G and Bonas U 1999. HrpG controls T3SS gene expression in *Xanthomonas*. *Molecular Microbiology* 33: 1179–1189.

White FF, Potnis N, Jones JB and Koebnik R 2009. The Type III effectors of *Xanthomonas*. *Molecular Plant Pathology* 10: 749–766.

Wu L, Wu H, Chen L, Yu X, Borriss R and Gao X 2015a. Difficidin and bacilysin from *Bacillus amyloliquefaciens* FZB42 have antibacterial activity against *Xanthomonas oryzae* rice pathogens. *Sci Rep.* 5: 12975. https://doi.org/10.1038/srep12975.

Wu LB, Eom JS, Isoda R, Li C, Char SN, Luo D, Schepler-Luu V, Nakamura M, Yang B and Frommer WB 2022. *OsSWEET11b*, a potential sixth leaf blight susceptibility gene involved in sugar transport-dependent male fertility. *New Phytologist* 234: 975–989.

Wu X, Kong D, Song S, Liu L, Kuang H and Xu C 2015b. Development of sandwich ELISA and immunochromatographic strip methods for the detection of *Xanthomonas oryzae* pv. *oryzae*. *Anal Methods*. 7: 6190–6197.

Wu X, Li X, Xu C and Wang S 2008. Fine genetic mapping of *xa24*, a recessive gene for resistance against *Xanthomonas oryzae* pv. *oryzae* in rice. *Theoretical and Applied Genetics* 118: 185–191.

Wu Y, Lee SK, Yoo Y, Wei J, Kwon SY, Lee SW, Jeon JS and An G 2018. Rice transcription factor OsDOF11 modulates sugar transport by promoting expression of sucrose transporter and *SWEET* genes. *Molecular Plant* 11: 833–845.

Xe S, Zang H, Wu H, Uddin Rajer F and Gao X 2018. Antibacterial effects of volatiles produced by *Bacillus* strain D13 against *Xanthomonas oryzae* pv. *oryzae*. *Mol Plant Pathol.* 19: 49–58. https://doi.org/10.1111/mpp.12494.

Xing J, Zhang D, Yin F, Zhong Q, Wang B, Xiao S, Ke X, Wang L, Zhang Y, Zhao C, Lu Y, Chen L, Cheng Z and Chen L 2021. Identification and fine-mapping of a new bacterial blight resistance gene, *Xa47(t)*, in G252, an introgression line of Yuanjiang common wild rice (*Oryza rufipogon*). *Plant Disease* 105(12): 4106–4112. https://doi.org/10.1094/PDIS-05-21-0939-RE.

Xu X, Li Y, Xu Z, Yan J, Wang Y, Wang Y, Cheng G, Zou L and Chen G 2022a. TALE induced immunity against the bacterial blight pathogen *Xanthomonas oryzae* pv. *oryzae* in rice. *Phytopathology Research* 4: 47.

Xu X, Xu Z, Ma W, Haq F, Li Y, Shah SMA, Zhu B, Zhu C, Zou L and Chen G 2021. TALE-triggered and iTALE-suppressed Xa1-mediated resistance to bacterial blight is independent of rice transcription factor subunits OsTFIIAγ1 or OsTFIIAγ5. *J Exp Bot.* 72(8): 3249–3262. https://doi.org/10.1093/jxb/erab054.

Xu Y, Zhu XF, Zhou MJ, et al 2010. Status of streptomycin resistance development in *Xanthomonas oryzae* pv. *oryzae* and *Xanthomonas oryzae* pv. *oryzicola* in China and their resistance characters. *J Phytopathol.* 158: 601–608.

Xu Z, Xu X, Wang Y, Liu L, Li Y, Yang Y, Liu L, Zou L and Chen G 2022b. A varied AvrXa23-like TALE enables the bacterial blight pathogen to avoid being trapped by *Xa23* resistance gene in rice. *J Adv Res.* https://doi.org/10.1016/j.jare.2022.01.007.

Xue N 1973. Application of streptomycin in plant protection. *Bull Agric Sci Technol.* (2): 21.

Xue YH 2002. Control effect of several biotic pesticides against diseased rice. *Jiangxi Plant Protection* 25: 68–71.

Yan Y, Wang Y, Yang X, Fang Y, Cheng G, Zou L and Chen G 2022. The MinCDE cell division system participates in the regulation of Type III secretion system (T3SS) genes, bacterial virulence, and motility in *Xanthomonas oryzae* pv. oryzae. *Microorganisms* 10(8): 1549.

Yang B and White FF 2004. Diverse members of the AvrBs3/PthA family of type III effectors are major virulence determinants in bacterial blight disease of rice. *Molecular Plant–Microbe Interactions* 17: 1192–1200.

Yasmin S, Hafeez FY, Mirza MS, Rasul M, Arshad HM, Zubair M and Iqbal M 2017. Biocontrol of bacterial leaf blight of rice and profiling of secondary metabolites produced by rhizospheric *Pseudomonas aeruginosa* BRp3. *Front Microbiol.* 8: 895 https://doi.org/10.3389/fmicb.2017.01895.

Yoshihisa A, Yoshimura S, Shimizu M, Sato S, Matsuno S, Mine A, Yamaguchi K and Kawasaki T 2022. The rice OsERF101 transcription factor regulates the NLR Xa1-mediated immunity induced by perception of TAL effectors. *New Phytologist* 236: 1441–1454. https://doi.10.1111/nph.18439.

Yoshimura A, Mew TW, Khush GS and Omura T 1983. Inheritance of resistance to bacterial blight in rice cultivar Cas 209. *Phytopathology* 73(10): 1409– 1412.

Yoshimura S, Yamanouchi U, Katayose Y, Toki S, Wang ZX, Kono I, Kurata N, Yano M, Iwata N and Sasaki T 1998. Expression of *Xa1*, a bacterial blight-resistance gene in rice, is induced by bacterial inoculation. *Proceedings of the National Academy of Sciences* 95: 1663–1668.

Yoshimura S, Yoshimura A, Iwata N, McCouch SR, Abenes ML, Baraoidan MR, Mew TW and Nelson RJ 1995. Tagging and combining bacterial blight resistance genes in rice using RAPD and RFLP markers. *Molecular Breeding* 1: 375–387.

Yoshimura S, Yoshimura A, Saito A, Kishimoto N, Kawase M, Yano M, Nkagahra M, Ogawa T and Iwata N 1992. RFLP analysis of introgressed chromosomal segments in three near-isogenic lines of rice for bacterial blight resistance genes, *Xa-1, Xa-3* and *Xa-4*. *The Japanese Journal of Genetics* 67: 29–37.

Young JM, Dye DW, Bradbury JF, Panagopoulos CG and Robbs CF 1978. A proposed nomenclature and classification for plant pathogenic bacteria. *New Zealand Journal of Agricultural Research* 21(1): 153–177.

Yu X, Liang X, Liu K, Dong W, Wang J and Zhou MG 2015. The *thiG* gene is required for full virulence of *Xanthomonas oryzae* pv. *oryzae* by preventing cell aggregation. *PLoS One* 10: e0134237.

Yuan M and Wang S 2013. Rice MtN3/saliva/SWEET family genes and their homologs in cellular organisms. *Molecular Plant.* 6: 665–674.

Yugander A, Sundaram RM, Ladhalakshmi D, Hajira SK, Prakasam V, Prasad MS, Sheshu M, Ravindra Babu V and Laha GS 2017. Virulence profiling of *Xanthomonas oryzae* pv. *oryzae* isolates, causing bacterial blight of rice in India. *European Journal of Plant Pathology* 149: 171–191. https://doi.org/10.1007/s10658-017-1176-y.

Zaka A, Grande G, Coronejo T, Quibod IL, Chen CW, Chang SJ, Szurek B, Muhammad A, Cruz CV and Oliva R 2018. Natural variations in the promoter of OsSWEET13 and OsSWEET14 expand the range of resistance against *Xanthomonas oryzae* pv. *oryzae. PloS One* 13: e0203711.

Zhai W, Wang W, Zhou Y, Li X, Zheng X, Wang S and Zhu L 2004. Breeding bacterial blight-resistant hybrid rice with the *Xa21* gene derived from wild rice *(Oryza longistaminata). Theoretical and Applied Genetics* 108: 105–112.

Zhang F, Zhuo DL, Zhang F, Huang LY, Wang WS, Xu JL, Vera Cruz C, Li ZK and Zhou YL 2014. *Xa39*, a novel dominant gene conferring broad spectrum resistance to *Xanthomonas oryzae* pv. *oryzae* in rice. *Plant Pathology* 64: 568–575. https://doi.org/10.1111/ppa.12283.

Zhang GQ, Angeles ER, Abenes MLP, Khush GS and Huang N 1996. RAPD and RFLP mapping of the bacterial blight resistance gene *xa-13* in rice. *Theoretical and Applied Genetics* 93(1–2): 65–70.

Zhang J, Yin Z and White F 2015. TAL effectors and the executor *R* genes in rice–Xanthomonas interactions. *Molecular Plant Pathology* 16: 775–785.

Zhang J, Yuan M, Chen L, Xie X, Yin Z and Wang G 2012. Genome-wide association study of rice bacterial blight resistance identifies new QTLs. *Rice* 5: 34–46. https://doi.org/10.1007/s12284-012-9086-2.

Zhang Q, Lin SC, Zhao BY, Wang CL, Yang WC, Zhou YL, Li DY, Chen CB and Zhu LH 1998. Identification and tagging a new gene for resistance to bacterial blight (*Xanthomonas oryzae* pv. *oryzae*) from *O. rufipogon. Rice Genet Newsl.* 15: 138–142.

Zhang Y, Li T, Liu Y, Li X, Zhang C, Feng Z, Peng X, Li Z, Qin S and Xing K 2019. Volatile organic compounds produced by *Pseudomonas chlororaphis* subsp. *aureofaciens* SPS-41 as biological fumigants to control *Ceratocystis fimbriata* in postharvest sweet potatoes. *J Agri Food Chem.* 67: 3702–3710. https://doi.org/10.1021/acs.jafc.9b00289.

Zhao S, White DG, Ge B, Ayers S, Friedman S, English L, Wagner D, Gaines S and Meng J 2001. Identification and characterization of integron-mediated antibiotic resistance among Shiga toxin-producing *Escherichia coli* isolates. *Applied and Environmental Microbiology* 67(4): 1558–1564. https://doi.org/10.1128/AEM.67.4.1558-1564.2001.

Zheng CK, Chun-Lian W, Yuan-Jie YU, Liang YT and Kai-Jun Z 2009. Identification and molecular mapping of *Xa32 (t)*, a novel resistance gene for bacterial blight (*Xanthomonas oryzae* pv. *oryzae*) in rice. *Acta Agronomica Sinica* 35: 1173–1180.

Zhou J, Peng Z, Long J, Sosso D, Liu B, Eom J S, Huang S, Liu S, Vera Cruz C, Frommer WB White FF and Yang B 2015. Gene targeting by the TAL effector PthXo2 reveals cryptic resistance gene for bacterial blight of rice. *The Plant Journal* 82: 632–643.

Zhu MJ, Ye DN, Zhang ZL, Liu GQ and Wu XF 1992. Test on control effect of new medicament to principal diseases of rice. *Plant Protection* 18: 26–27.

Zou L, Li YR, Pan X, Xiang X, Li C and Xu L 2012. The HrpG–HrpX regulatory cascade in *Xanthomonas* responds to plant-associated cues. *Applied and Environmental Microbiology* 78: 203–212.

Index

Note: Page numbers in *italics* indicate a figure and page numbers in **bold** indicate a table on the corresponding page.

D

E

F

G

H

I

K

L

U

V

W

X

Y

Z

For Product Safety Concerns and Information please contact our EU representative GPSR@taylorandfrancis.com Taylor & Francis Verlag GmbH, Kaufingerstraße 24, 80331 München, Germany